THE LIBRARY
ST. MARY'S COLLEGE OF MARYLAND
ST. MARY'S CITY, MARYLAND 20686

73442

A

VOYAGE OF DISCOVERY

TO THE

Strait of Magellan

MILFORD HOUSE
BOSTON

Library of Congress Cataloging in Publication Data

[Vargas y Ponce, José de] 1760-1821.
    A voyage of discovery to the Strait of Magellan.

    Compiled from the journals of Dionisio Alcalá Galiano and Alejandro Belmonte, lieutenants of Córdoba.
    A translation of parts of the author's Relación del último viage al estrecho de Magallanes de la fragata de S. M. Santa Maria de la Cabeza en los años de 1785 y 1786; published anonymously in Madrid, 1788.
    Attributed to J. Vargas y Ponce. Cf. Medina, Bibl. hisp.-chilena.
    Reprint of the 1820 ed. printed for R. Phillips, London.
    1. Magellan, Strait of. 2. Patagonia--Description and travel. 3. Spain--Exploring expeditions. I. Córdoba y Laso, Antonia de, d. 1811. II. Alcalá Galiano, Dionisio, 1760-1805. III. Belmonte, Alejandro. IV. Title.
F3191.V3 1973   910'.41   73-4993
ISBN 0-87821-077-6

This Milford House book is an unabridged republication of the edition of 1820.
Published in 1973 by Milford House Inc.
85 Newbury Street, Boston, Massachusetts
Library of Congress Catalogue Card Number: 73-4993
International Standard Book Number: 0-87821-077-6
Printed in the United States of America

A

VOYAGE OF DISCOVERY

TO THE

Strait of Magellan:

WITH AN ACCOUNT OF THE

MANNERS AND CUSTOMS OF THE INHABITANTS;

AND OF THE

Natural Productions of Patagonia.

UNDERTAKEN, BY ORDER OF THE KING OF SPAIN,

By ADMIRAL DON A. DE CORDOVA

Of the Royal Spanish Marine.

TRANSLATED FROM THE SPANISH.

LONDON:

PRINTED FOR SIR RICHARD PHILLIPS AND Co.

BRIDE COURT, BRIDGE STREET; AND TO BE HAD OF ALL BOOKSELLERS.

Entered at Stationers' Hall.

Printed by J. and C. Adlard,
23, Bartholomew-Close.

CONTENTS.

PART I.

	Page.
SECT. I. Preparation for the Expedition	1
II. Voyage from Cadiz to the Strait of Magellan	4
III. Occurrences in the Navigation of the Strait	12
IV. Return to Cadiz	51

PART II.

SECT. I. Description of the Strait of Magellan.—Division of the Country into high and low.—Temperature: Qualities of the Soil.—Productions of the Strait: Herbs, Plants, Flowers, Shrubs, and Trees.—Description of the Quadrupeds, Birds, Fishes, and Insects 64

II. Of the Inhabitants of the Strait of Magellan 84

III. Of the Indians of the Strait of Magellan 94

LINEAL MEASURES,

English, French, and Spanish.

                                              Eng. Inches.

THE most accurate standard of the yard, is that called the Parliamentary Standard, made by Bird in 1758, equal to .... 36·00023

On a measurement by the late Dr. Maskelyne, astronomer-royal, of two French standard toises, exhibited by M. de la Lande in London, the one was found to be equal to .... 76·732
And the other to .... 76·736

The medium 76·734 English inches for the toise of 6 feet, or 72 inches French, is adopted by Biot in his *Astronomie Physique*, 1810. According to this medium proportion, one inch, foot, fathom, or toise, English, is to one inch, &c. French as 1 : 1·06575 *precisely*. Or, making the French measure the unit, the English measure will be 0·938 of that unit *very nearly*.

In the end of 1801, Professor Pictet, of Geneva, carried over from London to Paris a brass scale, of 49 inches in length, made by Troughton. On a most accurate comparison of that scale, by Members of the Institute, with the platina and iron standard *mètres*, the *mètre* was found to be (at the the temperature of melting ice) equal to .... 39·38272
Which, reduced to the English rule of comparison, 62° of Fahrenheit, is 39·371

The *mètre* being the ten-millionth part of a quadrant of the meridian, a sexagesimal degree of a great circle must be equal to 57·012 French toises, or 60,760·54 English fathoms, equal to English miles .... 69·046

The late eminent nautical mathematician Captain Mendoza y Rios furnished this rule for converting Spanish into English measures, and *vice versa*.

From the logarithm of the number of Spanish feet, fathoms, &c. subtract the constant logarithm 0·0384544, and the remainder will be the logarithm of the corresponding number of English feet, fathoms, E. in. &c.
    Hence the Spanish foot is equal to .... 10·98313
    ———————— *vara*, or yard, equal to .... 32·94954
    ———————— *braza*, or fathom, equal to .... 65·89908
And 12 Spanish brazas are within a small trifle of 11 English faths.

N.B.—The French *brasse* contains only 5 French feet, and is therefore only equal to 63·945 English inches, equal to 5 feet and nearly 4 inches English.

PREFACE

OF

THE TRANSLATOR.

" CAN any good thing come out of Nazareth?" is a supercilious query but too frequently directed against various regions of the earth, particularly against bigotted and enslaved SPAIN. And, indeed, were our opinion of what *may* be extracted from that devoted country to be formed by estimating the magnitude, or the value of the information lately brought home from it, even by our friendly armies, after a long abode, the answer to the insolent query above-mentioned would justly be,— " Nothing."

Having resided for some time in Spain, and having occasionally, since my return, directed my attention towards Spanish literature, to that query I should feel myself warranted to give a very different answer.— To mention one branch of scientific literature only,— I mean what relates to maritime information respecting her dominions, European and American, *Spain has done more, and in a more systematic manner, than all the other Powers of Christendom put together.* She has surveyed, by the most able operators, furnished with the best instruments and instructions which England and France could produce, the whole sea coast of those vast regions; and the results she has, with a manly liberality without example, laid before the public. The great *Coasting Pilot* of Spain (including Portugal) was published by the authority of the Government in 1787, accompanied by a set of charts, illustrated by plans of all the principal harbours and naval stations, in the fullest detail; accompanied by views of the coast, the whole executed in a style far superior to any work of the kind which France or England yet possesses; except, perhaps, the Atlantic Neptune of Governor De Barres. This Coasting Pilot, forming a good quarto volume, I translated several years ago for Mr. Faden; and it now forms one of the standard treatises attached to every vessel belonging to the national navy.

Since then I have translated from the Spanish " An Account of a Voyage from Cadiz to the Strait of Magellan." This voyage, performed under the command of Admiral Don Antonio de Cordova, in a frigate, was projected with the humane view of setting at rest for ever the long-agitated question, whether mariners ought or ought not, on any account, to attempt a passage to the Pacific Ocean through that Strait.

The translation was first suggested to me by the late

ill-fated General Miranda; and, in the execution of it, I was encouraged and aided by the late first-rate nautical mathematician, Captain Mendoza y Rios, of the Spanish navy, but long domiciliated in Britain. That the subject is not only important, curious, and entertaining, in itself, and valuable, as filling up a chasm in our knowledge of the globe, will be seen from the annexed Contents; and, excepting the filling-up of chasms, little probability remains of discoveries of importance being now made. In the course of the voyages out from Spain to the Strait of Magellan, and home again, and in the perilous traverses of the Strait itself, various important particulars are finally settled by the able commander. The description of the countries bordering on the Strait, that is, of *Patagonia* and *Tierra del Fuego*; of the productions, animal and vegetable; of the famous gigantic race of Patagonians, in the work reduced to a rational standard;—these, and other matters, are well entitled to notice and to confidence.

The authority of the whole is not to be questioned. Its importance to mariners, particularly in the present eventful crisis of South American affairs, and of our intimate connections and frequent intercourse with the western coasts of the Spanish possessions (to say nothing of our growing navigation to New Holland and the whole Pacific), is self-evident;—the notices respecting the productions of the South American promontory, come forward with an air of genuineness sufficient to command our belief, and may in time lead to the discovery of other objects of singular value.

This voyage is calculated to be eminently useful at this time, by settling for ever the question in regard to the eligibility of making the passage of this Strait in a voyage to the Great Pacific:—a question which it decides in the NEGATIVE on the clearest evidence of facts. But, in arriving at this important conclusion, it has added to our stock of geographical knowledge, and brought us acquainted with large tracts, hitherto but imperfectly explored and studied.

ACCOUNT
OF A LATE
VOYAGE OF DISCOVERY
TO THE
Strait of Magellan.

THE King of Spain, having some time ago resolved to dispatch a vessel, for the purpose of examining and making an accurate draught of the Strait of Magellan, in South America, orders were given to the general of the fleet to select a frigate for the enterprize, who, in consequence, made choice of the Santa Maria de la Cabeza of thirty-six guns, whose good qualities were already well known. For several reasons, it was thought proper that she should not be sheathed with copper. She was built on the modern French model; which, it was believed, could not fail to render her fit for any kind of navigation; and the experience of this voyage confirmed that opinion: as her good qualities, on many occasions, delivered the officers and men from dangers, to all appearance inevitable.

The chief command of the frigate, and of the whole expedition, was conferred on Don Antonio de Cordova, of the Royal Navy, with leave for him to choose his officers, seamen, and marines; as also to carry out a second captain and two other officers, particularly conversant with astronomical observations.

In consequence of this permission, the commander made choice of Don Ferdinand de Miera to be his second captain; and it happened fortunately that Brigadier Don Vincent Tofiño was at that time in Cadiz, together with the officers of that department of the Spanish navy, who for two years past had been employed in the construction of a hydrographic atlas, or sea-chart, and coasting-pilot, of Spain: he requested to have two of these gentlemen, who, being thoroughly instructed in astronomical observations, and expert in the use of all the instruments necessary for this expedition, might assist him in the execution of his commission.

He therefore selected D. Dionisius Alcalá Galiano and D. Alexander Belmonte, lieutenants in the navy, who, without in the least excusing themselves from the strict performance of the ordinary duty of the ship, took the charge of all observations, astronomical and geographical; as also of the marine watches or time-pieces, and other instruments which were put on-board, as requisite for the due performance of the expedition.

Of this charge they acquitted themselves, to the entire satisfaction of the commander: and the following account of the voyage, together with the directions and instructions for navigating the Strait of Magellan, entirely drawn-up by these officers, is almost wholly compiled from their journals, formed and kept with the greatest skill and attention, and well deserving to be used as models for other voyages of a similar nature.

D. Antonio de Córdova also appointed D. Joachin Camacho to be his chief pilot, or master; and to him entrusted the operations necessary for taking separate draughts of the harbours, bays, and roads of the strait, and which we can affirm to have been executed with all the care and precision to be expected from the eminent knowledge and experience of that officer.

The following List shows the exact state of the whole officers and crew of the frigate, as she sailed from Cadiz:

Commandant.	Don Antonio de Córdova.
Second.	Don Ferdinand de Miera.
Lieutenants.	Don Miguel de Zapiain.
	Don Tello Mantilla.
	Don Dionysius Alcalá Galiano.
	Don Alexander Belmonte.
Midshipmen.	Don Pedro de Mesa.
	Don Joachin Blanco.
	Don Francisco Villegas.
	Don Philip Perez de Acevedo.
	Don Joachin Fernandez Salvador.
Officers of Marines.	Don Eugenio de Cardenas.
	Don Remigio Bobadilla.
Chaplains.	Don Julian Martiano.
	Don Joseph Riquero.
Surgeons.	Don Juan Luis Sanchez.
	Don Bartholomew de Rivas.
Pilots, or Masters.	Don Joachin Camacho, 1st.
	Don Antonio Castellanos, 2d.
Pilotines, or Mates.	Don Antonio Rico.
	Don Antonio Castro.
	Don Pedro Sanchez.

As the season was now far advanced for the nature of this voyage, (17th Sept.) no time was lost in getting ready the frigate; so that, on the 27th of the same month, she came out of the Dock-yard, completely equipped, with provisions for eight months, and wood and water for five; and, as the fortunate conclusion of all sea-expeditions must, in great measure, depend on the health and comfort of the seamen, it was the commander's principal care to use every precaution for that effect.

He therefore sent on-board, not only an ample supply of additional warm clothing for the crew, but also all such medicines, and other preservatives, as the experience of former navigators had shown to be essentially useful.

Whilst the commander was thus employed, the two officers, charged with the nautical observations, carried on-board two marine watches or time-pieces, (Nos. 15 and 16 of Berthoud,) which belonged to the Observatory of Cadiz, and also No. 71 of Arnold, a small one belonging to Lieut. Belmonte; and, by means of observations on-shore, and signals on-board the frigate, began to ascertain their several rates of motion, and to form the corresponding tables.

At the same time was carried on-board a complete collection of instruments, chosen from those procured by order of His Catholic Majesty in London, under the inspection of Mr. Jacinto Magellan, and from the hands of the most eminent artists of England. In removing them from the Observatory to the ship, the glass tube of the marine barometer suddenly burst, without its having received any blow or other injury,—so that the mercury was lost,—occasioned probably by the mercury having been but imperfectly purified, or perhaps not in the just quantity; so that we were under the necessity of proceeding on our voyage without that most valuable but delicate instrument.

Amongst these instruments were the best English quintants and sextants that could be procured: for if, by the aid of the time-pieces, we hoped to be able to ascertain every day the errors occasioned in our reckoning, by its own unavoidable uncertainty, as well as by the setting of tides and currents; so, by constant observations of the distance of the moon from the sun or stars, performed with these excellent instruments of reflection, by two distinct observers, it was to be expected, as it was afterwards experienced, that in each lunation we might discover what confidence ought to be placed in these time-pieces, supposing them to preserve one uniform motion: and, from such comparisons and calculations, we made no doubt of obtaining satisfactory results.

Lastly, as every information that could be collected respecting the object of the voyage must be useful, either as furnishing advice to be followed, or as pointing out what was to be avoided, the several officers of the frigate made it their business to draw together the accounts of all expeditions to the Strait of Magellan which had been published in the different parts of Europe; in which they were zealously assisted by Captain Don Alexander Malespina, of the department of Cadiz; who not only parted with his own collection of Voyages, but used every endeavour to procure from other persons such books as he himself did not possess.

All these preparations being concluded, and the commander satisfied that he now had on-board every thing requisite for the proper fulfilment of the enterprise, he ordered the vessel to be completely cleared on the 9th of October, and to be in perfect readiness to put to sea on the following morning.

SECTION II.

*Voyage from Cadiz to the Strait of Magellan.**

HAVING received from the Captain-general of the Fleet our final instructions and orders, authorising us to undertake the expedition, we set sail from Cadiz, at break of day, on Sunday the 9th of October, with an easy wind from the land, and an ebbing tide. The little wind we had being variable, we made but small progress; so that at night we were still within sight of the town. This weather lasted all night: however, in the morning early we had a distant view of Cape Spartel, on the coast of Africa. By observations of our latitude at mid-day, and from the longitude indicated by the time-pieces, as well as from our position according to the sea-chart, we found that the currents had carried us towards the south-east quarter,—a thing very common in these parts, on account of their general setting into the Strait of Gibraltar. In the evening we directed our course for the Canary Islands, which we reached without meeting with any thing remarkable on the passage.

The time-pieces, No. 16 of Berthoud and No. 71 of Arnold, kept an uniform pace: but No. 15 of the former artist gave room, from our first setting out, to suspect its accuracy; as its movements differed continually from those of the other two.

During the passage to the Canaries, no considerable setting of the waters was perceived; the errors occasioned by them in our reckoning mutually counterbalancing each other.

At day-break of the 16th, we discovered the islands Grand

---

* The name in the Spanish language is *Magallanes*, and pronounced nearly thus in English—*Magalyanes*, the third syllable being accented. The Italians spell it *Magaglianes*, which is equivalent to the former.

Canary and Teneriffe. At 7½ hours A.M. we took observations of the sun's altitude, in order to determine our longitude; and at the same time found the Pic of Teyde, in Teneriffe, bearing from us W. 3° S. and the W. point of Grand Canary, S. 14° W. from which observation we ascertained our place on the chart, making use of that of M. Verdun de la Crenne: and the latitude so pointed out agreed with that observed the following noon; when allowances had been made for a reckoning carefully kept during the interval of four-and-a-half hours, and for the setting of the waters to the southward; which, by observing the bearings of the island, we found had been the case. We were therefore at that time, according to the above-mentioned chart, in N. lat. 28° 18′ 20″, and in long. W. from Cadiz, 9° 28′ 48″. The time-piece, No. 16 of Berthoud, indicated 2′ 7″, and No. 71 of Arnold 3′ 4″ more to the westward: differences so inconsiderable on a course of such extent, as to give great hopes of reaping the highest advantages from those machines; whilst we entirely abandoned No. 15 of Berthoud, of whose accuracy we had all along been very suspicious. The errors in our ship's reckoning had been so compensated one by another, that it differed only a few minutes from the place laid down on the chart.*

All this day we continued under sail, with light winds, between the Grand Canary and Teneriffe, whose peak disappeared to us soon after midnight; nor did we see it any more on the following day, on account of the fresh weather. With tolerably fresh breezes we stood on to the S.S.W. that we might the sooner get out of the calms so often met with to leeward of the Canaries.

---

* It is necessary here to remark, that all the bearings mentioned in this account are corrected for the variation; and that the longitude is counted from the meridian of the Royal Marine Observatory at Cadiz 6° 16′ W. from Greenwich.

In the account of the last Voyage of the celebrated Captain Cook, an attempt is made to prove, that the position of the Road of Santa Cruz, in Teneriffe, is 14′ 30″ to the westward of that assigned by Captain Don Joseph Varela; who was employed, with Messrs. Verdun de la Crenne and Borda, in constructing the above chart. The observations of the Spanish astronomer must be preferable to those of Captain Cook, as being absolute, and independent of any errors occasioned by a time-piece. However, as much pains are taken in the above voyage to show that Varela was mistaken, we will here compare two positions, as laid down by Cook, with those assigned by two distinct astronomers; which, joined to Cook's remark on the Cape of Good Hope, which he lays down 8′ 25″ to the W. of the truth, will demonstrate that his time-piece went out of England with a positive error; or that, in a few days, it contracted that error.

Isle of Ouessant (Ushant), according to Cook W. from Paris	7°	37′	37″
Ditto, according to M. M. Verdun and Borda	7	24	30
Cook to the westward	0	13	04
Cape Finisterre, according to Cook, W. from Cadiz	3	12	00
Ditto, according to Don Vicente Tofino	2	55	54
Cook to the westward	0	16	06

Although it is the ordinary practice to cross the Equinoctial Line about the meridian of Teneriffe, (10° 22' W. from Cadiz;) yet, in order to avoid the calms which almost constantly prevail to the eastward of that point of longitude, we resolved to cross it more to the westward, viz. from 14° to 15° W. from Cadiz, and accordingly directed our course to that quarter.

As soon as we had passed the Tropic of Cancer, we began to distribute to every man of the crew a large plate of *gaspacho*, or spiced salad, by the use of which, and by regularly sprinkling vinegar and brine between the decks, we succeeded in maintaining them stout and healthy.

On the 24th at night, the breezes for the first time failed us, in N. latitude 11°. Nothing else noticeable having occurred since we lost sight of the Canaries.

A bright unclouded sky; winds regular and moderate, which temper the intense heat of the sun's rays; and a sea rarely agitated to excess: these circumstances have procured to this part of the Ocean the appellation of the Ladies' Bay. The sight we enjoyed every day of flying-fish, tunnies, and bonitos, diversified a little the monotony of this scene. We also had the company of a number of land birds, which, driven by the violence of the winds more than 100 leagues from their native country, hoped to escape unavoidable destruction in the water, by voluntarily delivering themselves into the hands of men.

About noon on the 25th, breezes again sprung up from E. and E.N.E., which, though feeble, still gave us hopes of speedily passing the Line. During the night we were assailed by a furious squall, which, leaving us no time to prepare for it, did some damage to our rigging; but, on the other hand, it furnished an opportunity of knowing the trim of our vessel; a circumstance which was afterwards of great consequence to us. The storm veered round by the N. to the W., and, after an hour's duration, it ceased; the heavens clearing up, which, during the squall, had a most unfavourable aspect.

On the morning of the 27th spoke with a Portuguese vessel, the Pez Medeo, out 38 days from Lisbon, and bound for Rio Janeiro in Brasil: her reckoning was very erroneous; and she sailed so ill, that we were not surprised at the small advance she had made in her voyage.

Calms and moderate winds continued successively till the 31st, when the latter became more constant from the N.E. and N.N.E.: our latitude being then about 7° 30' N.; and, on the 1st November, they changed to S.S.E.

We were now in longitude 14° W. from Cadiz. Standing on a tack to the eastward, with the wind at S.S.E. we made no advance in latitude, but on the contrary course towards the S.W. we gained a little southing, but still added more to our longitude, and were consequently departing widely from the meridian of Teneriffe, (10° 22′ W. from Cadiz;) the point in which, as we before said, it is customary to cross the Equinoctial.

As two or three ships, which have cut the line much to the W. of that meridian, have not been able afterwards to weather Cape St. Augustine, on the coast of Brasil, other navigators have been induced to keep as near to it as possible ; notwithstanding that experience has uniformly shewn, that the calms are more frequent and lasting, the sudden squalls and storms of lightning more violent and terrible, towards the coast of Africa, than they are towards the westward. In a climate of such a nature, the voyage is unavoidably much protracted ; the stock of water is exhausted ; and sickness, particularly the scurvy, is introduced into the ship, producing effects the most dreadful. The preservation of the health and lives of the ship's company ought to be the principal object of a commander's solicitude; not only because humanity requires it, but because it is the only way to attain the accomplishment of the expedition entrusted to him : and his care and attention ought to be more particularly exerted in a long voyage, in which little or no refreshment or relief can be expected. From these considerations, we resolved to follow such courses as would produce the greatest southing, in order to be freed from those latitudes so subject to calms, squalls, and sudden gusts of wind. We had besides observed that our frigate sailed remarkably well and close to the wind, so that we had reason to hope, if we should fall a little to the leeward, that we would soon recover a proper situation; especially as we were very confident in our methods for correcting errors in our reckoning, and in our calculations of our position with respect to the American coast.

The winds continued to blow from S.S.E. to S.E. with the sky cloudy and squally, as is generally the case in these parts. On the 8th we discovered a flaw in one of our top-masts about a yard long, and of considerable depth, which obliged us in future to proportion our sail to what we thought the mast would bear.

On the 9th November, at 10 P. M. we crossed the equinoctial line at 19° W. from Cadiz, (351° 22′ from the meridian of Teneriffe, counted easterly,) according to the time-piece No. 16, which coincided with No. 71, and both within half a degree of the longitude deduced from observations of the moon's distance from the sun. Our reckoning placed the frigate 54′

more easterly than the time-piece, and 1° 22' more easterly than the observations.

From repeated observations of the variation of the magnetic needle, both by amplitude and azimuth, we found the variation chart of M. Bellin, constructed in 1757, sufficiently exact in the neighbourhood of the line; but by no means so much to be depended on, when at some distance from it.

The discussions between Captain Cook and M. de Monier, on this subject, excited us to the inquiry: in which our results coincided with those of the former navigator; although we could not justly say of our needle as the captain says of his,—that it was excellent.

As soon as we had passed into S. latitude, the winds fixed themselves in the S.E. quarter, drawing however more to the E. than to the S.; which was also observed by Captain D. Joseph Varela, of the Spanish frigate Santa Catalina, who, in his voyage from Cadiz to Brasil, in 1777, being obliged to cross the Line far to the W. of the usual point, apprehended he should not be able to make Cape St. Augustine; but he found that, as he advanced in the southern hemisphere, and approached the coast of Brasil, the winds drew more to the E.; so that, although he had fallen so much to leeward, as to be within sight of the island of Fernando Norono, he yet weathered that cape without any difficulty. Cook experienced almost the same favourable change of the wind, as do all our vessels on their return from America to Europe; for which reasons it seems not improper that, in a voyage from Europe to South America, navigators should venture to trust a little to such a change; and therefore to cross the Equinoctial from 15° to 20° from Cadiz, provided the properties of their ships warrant them to make such attempts.

Captain Cook, in consequence of his own experience, lays it down, that from the meridian of Teneriffe to the W., the setting of the waters is constantly to the westward; and, on the E. side of the same meridian, they set to the eastward: that is to say, that they draw always towards the respective coasts of America and Africa. Our experience in this voyage confirmed the former part of the observation, which explains the variety in the accounts of different navigators respecting these currents. Some instances which we had occasion to consult also attest the truth of the remark: for example, Captain D. Philip Gonzalez, of the Spanish man-of-war the San Lorenzo, on his passage from Spain to Lima, having crossed the Line to the eastward of the meridian of Teneriffe, found, on his arrival on the coast of Patagonia, a great difference to the eastward between his reckoning and the true situation of that coast. He had ascertained his place by the chart of M. Bellin, the correctness of which, on that coast, we had reason to testify.

On the other hand, ships returning to Europe, which pass always to the westward of that meridian, find the errors in their accounts to be constantly to the westward of the truth. Some instances might also be mentioned of vessels which found little or no errors in their reckonings which could be attributed to the setting of the waters, because they had passed at or near the meridian of Teneriffe.

The breezes continued from the eastward until we had reached south latitude 18°, when we experienced their usual change, viz. that as we advanced in south latitude, they veered round more and more to E. and NE. which enabled us without any difficulty to pass Cape St. Augustine; and to cross without apprehension the parallel of the Abrojos, notwithstanding that the currents had set considerably to the westward, as we could easily perceive by our observations for ascertaining the longitude. In south latitude 22° 45', these winds failed us; and, which is worthy of notice, at the same time also ceased the setting of the currents towards the W. For two days we had winds from S. and SW., but they afterwards returned to the NE.

On the 21st November fell in with the ship Los Placeres, belonging to the new Company of the Philippine Islands, which sent her boat on-board of us as soon as we were within her reach. She had sailed from Cadiz on the 1st October for Lima, and had crossed the Equinoctial two days later, and two or three degrees more to the eastward, than we had done. Her reckoning, corrected for currents to the W., agreed with our account deduced from our observations of lunar distances by the time-pieces. We steered a little more easterly than she did; and, from our superiority in sailing, next day lost sight of her.

In south latitude 31° 45' the wind began to veer round by the N. and W. to the SW. quarter. Experience has shown, that, in these latitudes, and in the neighbourhood of the coast of Brasil, when the wind has been at NE. and draws round by the N. to the W., it never fails to arrive at the SW., from which point it blows with the greatest violence. These last winds are commonly called Pamperos, as proceeding from, or over, the immense plains or Pampas, adjacent to the River de la Plata in South America.

From latitude 13° 5' to latitude 41°, the longitude indicated by the time-pieces was constantly 2° or $2\frac{1}{2}°$ to the W. of that drawn from our reckoning; but after that parallel (41°,) in which the winds began to blow from the NW. and SW. quarters. These machines pointed out a considerable setting of the waters to the E.; insomuch, that from the 8th to the 13th December, in which time we crossed the parallels from 41° to 45°, there was an error in our reckoning of 1° 35' of longitude. There was likewise an error of 1° 32' of latitude; and both were evidently occasioned by the currents towards the E. and N. The drift of the sea in that direction first gave us notice of that which observations, made on purpose, afterwards confirmed.

A little before noon of the 11th December we felt a quick short motion of the vessel, which was repeated after a short interval, and was supposed to proceed from an earthquake, agreeably to the opinion of some of our officers, who had experienced the like on other occasions. On the following day died Alonzo Mateo, a seaman, with no other symptoms of illness than a violent colic, which carried him off in a few hours. The loss of one of our company was sensibly felt; however, we had the satisfaction to observe that all others continued to enjoy good health; and that the transition from a warm to a cold climate had made no considerable alteration on them.

From south latitude 43°, we directed our course to discover Cape Blaneo; but were prevented from seeing it by the westerly winds.

We cannot sufficiently recommend it to navigators in these seas, bound for the Strait of Magellan or the Pacific Ocean, to keep as close as possible to the Patagonian coast, in summer time; that is, in the months of November, December, January, and February, during which the cross-winds from the SE. are not known; otherwise, the winds which predominate from NW. to SW. will prevent them from doing it when it is absolutely necessary. If, in a passage to Lima, a vessel should pass much to the eastward of Cape de las Virgines, and in that latitude should meet with winds from SW., it would afterwards cost infinite labour to make States Island; and, consequently, it would be to enhance unnecessarily the difficulties and dangers of a voyage, which is already one of the most hazardous and severe that can be undertaken.

At length, on the evening of the 13th December, we came into soundings, with 60 fathoms fine brown sand. We were then distant from the land about 64 leagues, according to the chart of M. Bellin; our time-pieces and lunar observations shewing us to be in west longitude 55°, our latitude at the same time being 45° 35' 5" south.

From that position we judged a SSW. course to be the most advantageous, and accordingly steered to that quarter.

It is proper to remark in this place, that the soundings on the coast of Patagonia are by no means so regular as some navigators have imagined them to be. It is a very common notion, that the number of leagues of distance from the land is, in general, equal to the number of fathoms of water in the soundings; but, on the contrary, in approaching the coast, we sometimes found that the depth of water, instead of diminishing, increased: so that the true guide is the sight of the land, when there is generally from 18 to 24 fathoms, as many navigators have remarked.

There is not the smallest danger in approaching the Patagonian coast, which is every-where clean and safe, excepting only in south latitude 48° 34', where there is a shoal lying five or six leagues out from the shore. The two brothers Nodal, who examined these seas in 1619, and M. de Bougainville in 1764, have both fixed its position, agreeing very nearly as to its latitude; so that there can be no doubt as to its existence.

From the time that we came into soundings we saw a prodigious number of sea-fowl, as well as of whales, and sea-lions or seals, which abound all over these seas.

On the 18th December at sunset, at last, we had the happiness to obtain a sight of the land, five or six leagues off, in south latitude 51°; but the horizon being dim, we could not know precisely what part it was: only we concluded, from the latitude, that it was the bay at the mouth of the river Santa Cruz. We were then in 45 fathoms water, with brown and black sand. All night we stood on along the coast, steering S. and S. by W. sounding from time to time, from 48 to 48 fathoms, fine brown and black sand; and, in the morning, observed Cape de Barreras Blancas, called by Wallis and other foreigners Cape Fair Weather. According to observations at noon, this cape lies in south latitude 51° 31' 30"; and measuring the difference of meridians between it and Cape de las Virgines, afterwards correctly determined, its longitude is 62° 40' 30" W. from Cadiz.

To the southward of Cape Fair Weather we observed seven little hills a good way up from the shore, but very near one to another, and all about the same height, which, for these reasons, we called Los Frayles (the Friars).

At 11 A.M. we got within sight of the long-looked-for Cape de las Virgines. The wind had blown very strong all the day from the SW.; but at 3 P.M. it fell calm, when, considering that we might ride in safety under the shelter of the land, without being driven to leeward, should the wind become more violent and contrary, we resolved here to wait for a favourable opportunity for entering the Strait of Magellan, of which the above cape forms the N. and E. point; and therefore came to anchor in 28 fathoms fine sand, distant four leagues N. from that cape, and one league and a half from the Patagonian shore.

SECTION III.

Occurrences in the Navigation of the Strait.

No sooner had we let go the anchor, and hoisted out the boat, than, on the opposite shore, we perceived a number of fires lighted up, indicating that it was inhabited; and being extremely desirous of examining with our own eyes a country so variously described by different travellers, we rowed in to the part of the shore where the natives were assembled, and all on horseback, to receive us, showing however some apprehension of danger from us. We made signs of friendly intentions, presenting them with some trifles, and soon gained their confidence so far that they accepted our invitation to go on-board the frigate. The carelessness with which they left their companions on the shore, their horses, and other things which were not wanted, while they went on-board, gave a convincing proof of the probity and good faith which subsist among them: One of those that embarked spoke a few words of Spanish; from which circumstance we concluded that he must have had considerable intercourse with the Spanish settlements on the northern parts of the coast. He mentioned the names of several Spaniards, particularly Captain Antonio Viedma, and the pilot Bernardo Tafor. He wore a sort of cloak passed over his head, made of the cloth manufactured by our settlers in the province of Rio de la Plata. In short, his whole dress resembled that of the creoles of South America, with the addition only of a sort of cloak of the skins of the lama or guanaco stitched together, and precisely like those which are made for sale by the Indians of that province. His name he said was Francisco Xavier, and was of the ordinary size of men. He seemed to have complete authority over his companion, whose overgrown figure, being six feet eleven inches and a half Spanish measure in height, could not fail to excite much surprise in us inhabitants of the old continent. Francisco's attention seemed to be chiefly fixed on a sabre he wore, and which, after sundry applications, he suffered us to examine. It was in the scabbard, and on the blade were the following words, Por el Rey Carlos III. (for King Charles III.) They were both provided with spears and bows, arms well known in the province of Buenos Ayres.

That he might conciliate our goodwill, Francisco showed the greatest desire to gratify our curiosity. Having observed that we were surprised at the appearance of his companion, who had circular marks painted about his eyes, red in the upper part and black in the lower, he directed him to remove it, which he immediately did, wiping it off with the corner of his skin-cloak.

They both conducted themselves with great frankness and cordiality, showing that they entertained no apprehensions of injury from us. They smoked tobacco, they sat down to the table, making dexterous use of the knife and fork, and spoon; but they positively refused to drink any wine or brandy: and Francisco having been prevailed on to taste a little of the brandy, immediately rejected it, and warned his companion not to touch it, giving us to know, at the same time, that he was not ignorant of the hurtful effects of these liquors.

We provided beds for them, where they slept; and, in the morning, carried them on shore, seemingly much pleased with their excursion. On their landing, they signified to the boat's crew that they wished them to wait until they should bring a present of lama's flesh and skins; but having been ordered to return to the frigate without delay, they were obliged to decline the obliging proposal.

We continued in this anchorage till the 22d in the morning, having at times very fresh gales from the SW. quarter; at other times, it was quite calm: but this day the wind being brisk from N. and NW., we set sail for Cape de las Virgenes, keeping our launch a little way before us, to make signals of the depth of water. At three P.M. we got to the mouth of the strait; and at five P.M. finding the tide to be against us, we came to anchor in 28 fathoms coral bottom, being about one mile from the north shore, and two leagues within the strait.

At eight P.M., finding the tide to be on the turn, and the weather quite calm, we weighed anchor, in order to get into a better situation; and, being towed by the launch and another boat with the help of the tide, now favourable, we continued making way till eleven P.M., when we again came to anchor in 15 fathoms mud and sand.

We perceived no change in the currents until about $3\frac{1}{2}$ A.M. of the 23d, when we lay with our head to NW. by W., and at five A.M. their force was directed towards SE.

The Patagonians, who had remained all the preceding day on the shore opposite to the vessel, but had retired at sunset to some hollow broken grounds back from the sea, returned to the beach at break of day in greater numbers than before, all on horseback, and attended by their dogs. Many of them alighted, and began to dance and leap about, in token of their friendly dispositions.

As the safety of our vessel required that we should find more proper anchorage than that where we lay, we weighed at eight A.M. observing the tide to be on the turn, and steered for Possession Bay, sending on the launch before us, to make signals of the depth of water and quality of the bottom, that the frigate might be the less exposed to accidents. Having stood for two hours on a tack to the northward, till we got into seven fathoms fine brown sand, we began to put about for the southward; but the vessel, having little way, was so long in stays, that before she got round, we were in less than four fathoms water, and only a little more than a cable's length from the shore, where we heard distinctly the natives repeating the sounds of the seamen on board the frigate,—a strong proof of the acuteness of their hearing, and of the great manageableness of their organs of speech.

We continued, after this tacking, with easy winds from the W. to the SW. till half-past two P.M., when we began to lose ground; to avoid which, we came to anchor in 24 fathoms dark-coloured sand, Cape Possession then bearing W. 18° N. and the high land of Cape de las Virgenes N. 58° E. At half-past four P.M. the wind changed to the E. still easy; and although the wind was contrary, yet we hoped with that wind to make way against it; but, at six P.M. it again blew from the W.; nevertheless, we continued tacking till half-past eleven P.M., when we again dropped anchor in 24 fathoms fine sand and mud.

As the wind soon afterwards came to blow hard, we became alarmed for our launch, which was at some distance, and were obliged to ply a good deal to windward before she could get alongside. All this night we remained at anchor, the wind blowing fresh from WSW. to W. At nine in the morning we reefed our top-sails, to be ready to get under sail as soon as the current should cease to be against us, which we executed at noon; and continued making tacks and sounding all the way, carefully examining the bottom, till seven P.M., when we came to anchor in 16 fathoms brown sand, Cape Possession bearing W. 22° N. and the hill Del Dinero N. 48° W. The wind was constantly changing, so that it was still contrary on every tack we made for the space of seven hours; on which account we made very little way, not having advanced four leagues in five days of incessant exertion.

The current set to the SE. till half-past eleven P.M., when we rode to the wind, which came on with heavy squalls. The sky was cloudy and dark the whole night; so that we had the mortification of being prevented from observing the emersions of the 1st and 2d satellites of Jupiter, by which we would have been able to determine with accuracy the longitude of Cape de las Virgenes.

All the 25th we remained at anchor, with the wind fresh, accompanied by a heavy sea from SW.: the current was either at a stand, or setting towards ESE. We made several experiments to ascertain its rise and fall, but were never able to satisfy ourselves, the vessel being as it were in a kind of whirlpool, from the continual shifting of the wind and currents: upon the whole, however, we were of opinion that the rise and fall of the water are very inconsiderable.

On the following day we had but little wind, but the current continued to be almost always contrary to our course. In the evening, the boat went on shore, for the purpose of making some observations requisite for laying down the chart of the strait. The Patagonians, who for some days past had not appeared on the shore, but remained about a mile up the country, endeavouring to soften the rigour of the weather by keeping up large fires, now came down, in number about thirty, all men, to the place where the boat was to land, and received our people with every mark of friendship and goodwill. We presented to them some glass trinkets, making them sit down, that we might tie them round their necks with red ribbons; and our second captain gave to the tallest among them a small plate of metal, with his name and the date engraved on it. We invited them to come on-board the frigate, which they declined doing at that time, as night was coming on, when they regularly returned to their place of abode; but promised to pay us a visit on the morrow, or next sun, as they express it.

Having no kind of boats or ships themselves, they have consequently no names for them; they therefore called the frigate the great waggon or cart, and the boat the little waggon.

On this occasion we had again an opportunity of remarking the quiet temper and dispositions of these Patagonians, as well as their manners and stature. Their continuing in that part of the strait as long as we remained there, seemed to prove rather their desire to acquire some European trinkets, jewels in their eyes, than any wish for conversation or intercourse with us. The tribe consisted of 300 or 400 persons, all men and boys, for we saw not one woman among them, who no doubt remained at their habitations up the country, whither the men every night retired.

All requisite operations for taking a survey of that part of the strait being performed to our satisfaction, the boat returned to the frigate.

On the 27th the wind freshened from the SW. with terrible squalls, the sea also growing furious, so that the vessel, which had hitherto remained at single anchor, began to drive towards the land. In this situation it was resolved to let-go another anchor, that we might neither lose the ground we had gained, nor expose the launch, which was a little way a-head; and which it was equally impossible to take on-board as it was imprudent to abandon to her fate. The weather continued in this way till the 29th, at $4\frac{1}{2}$ P.M. when, after a furious gust of wind, one of our cables gave way, and the ship began to drive on the other anchor: we immediately got out a third, which brought her up. However, she had already driven so much, that the point of Miera remained behind us only about two miles off, and right across the direction of the wind. In this situation we had only the alternative, either to depend on the resistance of our two cables, or instantly to get under sail before, by driving a little farther, it should be utterly out of our power so to do. This last operation, in the present state of the vessel, considering that the violence of the wind would not admit us to set much sail, became equally difficult and dangerous. It was however determined to attempt it, even at the hazard of losing our launch. We therefore cut away the two anchors, and got under sail, executing the whole with such alacrity and success, that, by means of the current, and the good qualities of the vessel, we got safely out of this most imminent danger, although at the expense of the greatest sacrifices which in our circumstances could be made. In this most unfortunate event we lost three anchors, and more than four cables, which filled us with apprehensions for the final success of the enterprise. But a slight consideration of all circumstances will show, that the above steps were indispensable; and that the fury of the wind, which occasioned our heavy loss, was little to be expected, especially in a place where the greatest part of former travellers had anchored before us. We now ran out of the strait, and immediately stood out to the northward, until we got under the lee of the land, to take in our launch, in which we fortunately succeeded, but not without much pains, in which the boat suffered great damage, and was reduced to the worst condition. The winds continued all night from the westward with such fury, that we were able only to set the mizen stay-sail; so that in the morning we had been driven above sixteen leagues from the coast.

We had now remaining for performing our undertaking only

three anchors, with two small ones, which might be joined together so as to make one, with two whole cables and two pieces of others. Notwithstanding such slender provision of indispensable articles, it was still unanimously resolved, by all the officers on-board, to return to the Strait as soon as the wind should permit: and it having subsided considerably on the morning of the 31st, but still blowing from WSW. we bent our course to the southward; and, on the 1st of January, at break of day, we again came in sight of Cape de las Virgenes, distant about four leagues. At 8 A.M. we reached the mouth of the Strait, and having turned to windward the whole day to little advantage: in the evening, the west assuming again a very threatening aspect, we thought it to be most prudent once more to take refuge under the land to the northward of the same cape, there to wait for more favourable weather.

The wind abating a little on the 2d, we entered the Strait a third time; and, having tacked all the morning to little purpose, it fell calm about 1½ P.M. when the current being against us, we dropped one of our small anchors in 43 fa. fine sand and shells, Cape de las Virgenes, bearing N. 31° E. 4 leagues. At 3 P.M. the wind sprung up from SW., and the force of the current being weakened, we weighed; and, notwithstanding the approach of night, and the unfavourable appearance of the heavens, continued under-sail, preferring rather to turn up to windward, even in these dangerous parts, by night, than again to hazard the loss of more cables and anchors. The night was spent in tacking and sounding, putting about always when we came into 14, 15, or 16 fathoms water.

The remark seems to be well founded, that, in these parts, when the currents are very strong towards the E. but almost insensible in their course to the westward, it is a sign that the winds will be boisterous and squally from the W.; for, in this, our third attempt to enter the Strait, in which the winds had neither been constant nor violent from the westward, we found that the currents had been considerably in our favour. Notwithstanding that it is not advisable to turn to windward by night in such passages, yet, in the morning, we found that we had advanced more in the Strait than we had done in all the days preceding; and at 11 A.M. we had reached Cape Possession, although the wind had always blown from the westward. At 6 P.M. that cape bearing N. 25° E. and the hill called Aymon W. 54° N. it being quite calm, and the current setting SE. we anchored in 10 fathoms brown sand, and got out the boat, to sound in the neighbourhood of the vessel; and to discover the channel which led to the first pass, or Pass of Nuestra Señora de la Esperanza. At 6½ P.M. the current began to set to the SW. and, at 7¼, went at the rate of two miles in an hour in that direction. At 9 P.M. the boat returned, informing us, that, in steering to the WNW. there was a sufficiency of water, but that the entry of that pass had not been observed. At 9½ we got under-sail, in search of better anchorage; and at 10½ P.M. again anchored in 23 fathoms sand and stones, believing ourselves to be at no great distance from the pass. The greatest force of the current here, that we could discover, was about three miles an hour.

At 8 on the following morning, the current, which, since 2 A.M. had set towards the NE. having lost much of its force, we got under sail; and, at noon, being opposite to that pass, the current setting westerly, carried us even against the wind, until we reached the middle of the channel, when the wind falling, and the current beginning again to return towards the NE. we were obliged to come to anchor in 38 fathoms, lying very close to the north shore of the pass. We immediately sent the boat to land, in order to continue our geometrical operations, which returned at 8 P.M. every thing being performed to our satisfaction, and brought a great quantity of shell-fish. It was impossible to procure any lamas or guanacos, although many were seen at a distance; and three zorillos, which were shot, were obliged to be thrown into the sea, on account of the abominable smell of their water, which might be perceived at a great distance.

The strength of the current increased continually, until, at 7 P.M. it amounted to 7$\frac{5}{7}$ miles in an hour. Hence, if we suppose that its greatest force must be in the middle of the channel of this first pass, we may conclude it to be one of the most rapid on the face of the globe.

In the evening of this day ceased, for the first time, the reigning winds from the W. and for three hours together we had them very fresh from NE. and N. Notwithstanding the danger of night-navigation in these parts, we resolved to take advantage of this wind, as well as of the current, which would soon be in our favour. We therefore began to weigh, and, after some difficulty, got the anchor a-peak. Being now ready to set sail, in the hope of making some progress during the night, we were far from imagining that, in this place, fresh disasters awaited us. Neither the two capstans, nor any other contrivance we could think of, were able to raise the anchor, but all gave way, and the cable ran quite out to the end, made fast to the mainmast; so that it became now most difficult to set sail in the dark, and in a current running SW., first, at the rate of 4$\frac{3}{7}$ miles, and, afterwards, of 6 miles per hour. In this situation, our distress was extreme; for the vessel was running up the Strait at the mercy of the current. We therefore began to sound, and tried to take advantage of a small breeze from the SW. in order to keep, if possible, in the middle of the pass. At length, when we had got-in the cable, we found that the anchor was broken, the greatest part of it remaining in the sea.

How great this loss was to us, may easily be imagined, when it is considered, that few or no resources were to be found in this place, where the dangers to which we were continually exposed rendered cables and anchors of indispensable necessity. Former navigators had arrived at this part of the Strait of Magellan without sustaining any loss whatever, or running any other danger than what arose from the neighbourhood of the shores, all but very little known, united to the violence of winds almost constantly unfavourable. The wind at this time was very low; but the current hurried us along the middle of the channel, and at last brought us clear, and out at the west end of the pass. At 2 A.M. a light breeze arose from the W. with which, at $2\frac{1}{2}$, we reached the north shore, and came to anchor in the Bay of S. Gregorio, the cape of the same name bearing S. 88° W. about two leagues distance; having sailed seven leagues during the night, carried along almost the whole way by the force of the current.

At $9\frac{1}{2}$ A.M. we again weighed, to take advantage of the tide or current, now in our favour, but without any wind, till $11\frac{1}{2}$, when a small breeze from S.W. arose; and, after a number of tacks, we ran through the second pass of this strait; at $3\frac{1}{2}$ P.M. reached its west end; and used every effort to arrive at the anchorage of the Island St. Isabel, (Elizabeth,) notwithstanding that the wind now blew fresh from SW. But, the tide beginning to alter its direction, we were obliged to look out for a proper situation where we might wait for its return in our favour. As we were very unwilling to sacrifice the way we had now gained, we resolved to embrace the first opportunity for anchoring that should present itself on the north side of this pass: we at last come to anchor in a place that afforded some shelter, in five fathoms sand; but the water falling off speedily one fathom, we were obliged instantly to quit that place, knowing well the danger of such a situation; and were fortunate enough to be able to weigh the anchor quickly, although both wind and tide were against us. It was in this place that Sarmiento anchored, who, for a similar reason, was likewise obliged to abandon it. He called the bay Santa Susanna.

From on-board the frigate we observed a shoal, about two miles off, and a good way out from the north shore, which is not mentioned in any charts or accounts of former navigators. Soon after we got under-way, we discovered that it extended considerably in our direction; and, on the fall of a wave, our helm struck on a part of it, but without doing any material damage: we therefore crowded sail until we got into deep water; and, at night-fall, again returned to anchor in the Bay of St. Gregorio. The violence of the winds from W. and SW. kept us in this place till the morning of the 8th, during which time, we ascertained, by observations, the latitude and longitude of Cape St. Gregorio; but this day, the wind being fresh from the W. and the heavens clear, we again set sail, to continue our course to the westward. The cable which had served in this anchorage received considerable injury from the ship's bottom; whilst the tide, whose strength never exceeded four miles, was contrary to the wind.

This circumstance added not a little to the wants of our vessel, already too numerous: it was nevertheless determined, that we should still continue in the discharge of the commission with which we were intrusted.

We ran through the second pass of St. Simon with great ease, favoured by the wind, which veered round by the NW. to NNE., and steered for the channel between the Island St. Elizabeth and the little isles St. Magdalen and St. Martha, which is the most difficult passage in the whole Strait of Magellan, on account of the many reefs of rocks which run out from all these islands. In this situation fresh labours and dangers were not wanting to us: the wind drew suddenly round to WSW. and drove us over the reefs running out from St. Magdalen amongst the sea-weed, which is almost constantly found on the shoals and sank rocks of this strait, known to the Spanish seamen by the name of *cachiyullo*, but by naturalists called *fucus giganteus antarcticus*. The squalls becoming more furious than before, the heavens more obscured, the bottom more rocky and stony, and the depth of water rapidly diminishing even to four fathoms on each side of the vessel,—all this placed us in the most critical situation we had yet experienced. It was necessary to carry a press of sail, to prevent us from falling still farther on the shoal; nor could we continue long on any one tack, as the danger seemed to be alike on all hands; so that we were obliged to change our course repeatedly,—well understanding the dreadful consequences that must attend the loss of the frigate in such a situation. At last, fortunately, we got into deeper water on to 20 fathoms; and, although in following the same tack, we fell again near to another shoal, formerly observed by Captain Wallis, we also got out of that hazardous position, and soon discovered by the lead, that the depth was fast increasing. After this, we stood over to the westward, to get in with the north shore of the Strait, meeting frequently with furious blasts from the breaks in the mountains of the continent; and, towards night, came to anchor about one mile from the land, and one league north from the point of St. Maria; having received, in all this terrible day's work, but inconsiderable damage; and being enabled, from our own experience, to recommend it most earnestly to all navigators in these parts, to keep as close as they possibly can to the coast of the continent, that they may escape the dangers with which we had been so lately surrounded.

All the night the weather was cloudy, with showers of rain, or rather snow: the wind continued with most furious blasts, which, however, only effected one upper rigging, without reaching the hull of the vessel, which was protected by the excellent shelter of the land. The following day was employed in examining, with the boat, such parts of the coast as seemed to be interesting, and making proper observations for constructing the chart of the Strait,—a point, of which we never lost sight on-board the frigate, not even in the moment of greatest difficulty and danger,—sounding and making remarks as often as possible, even when our course, and the situation of the ship, did not require it.

The boat returned from this survey, bringing off five Indians who had been met with on the beach, whose nudity, loathsomeness, and stupidity, filled us with equal disgust and commiseration, naturally excited at the sight of such unfortunate objects, who exist in the utmost wretchedness, with which, however, they appear to be contented, doubtless from the very confined circle of their ideas; for, being ignorant of such things as they do not want, they are sensible of no privation. On their return to the shore, they joined their companions, who waited for them, and all together accompanied the frigate, by land, on her way to Puerto de la Hambre (Port Famine).

On the 10th of January we sailed for that harbour, where we anchored at 1 P.M. in 28 fathoms, after turning up against the winds from WNW. to WSW. the whole way. We earnestly recommend it to navigators to have their top-sails reefed in sailing near this coast: otherwise they may meet with very serious injury, from the sudden gusts of wind that break through the rugged openings of the mountains.

Port Famine is so called from the wretched remains of the settlement of San Felipe, visited by Cavendish, the English traveller, in 1587, at the time they were perishing with cold, hunger, and every other distress. This harbour furnished us with the means of repairing some part of our damages; of providing wood and water, fish, and some other articles of refreshment; of acquiring some knowledge of the inhabitants of that part of the Strait, and of placing the frigate in some degree of safety whilst the boats were engaged in the necessary examinations of both sides of the Strait. Here we completely re-fitted our launch, which had so much suffered in the entrance: we formed a stock to unite our two small anchors, adapting to them a Dutch fourteen-inch cable. We observed several meridian altitudes of the sun, to determine the latitude of the harbour; we examined the motions of our time-piece by observations of the sun's absolute altitude, seeing that the state of the heavens would not admit of our taking corresponding altitudes; and lastly, here the use of wild parsley, with which the shores abound, the opportunities of going on shore, the relief from the constant labour and anxiety which we had so long undergone, were of infinite service to the people of the frigate, whose steadiness, good-will, and general health, it was more easy to admire than to describe.

When we had finished our survey of the harbour and examined the adjacent coasts, the boat proceeded on other expeditions. We went along the shore as far as the west part of French Bay (Bay of St. Nicholas), making a draught of it as we went along: from thence we crossed over the Strait to the Land del Fuego, when we discovered an excellent bay and harbour, to which we gave the names of Bay of Valdés and Port Antonio, in honour of Sr Don Antonio de Valdés, which places were totally unnoticed by all former navigators.

The natives who had been on-board the frigate while we lay near Point St. Maria, as was before observed, followed us along the shore, accompanied by the rest of their tribe, amounting to twenty-three persons, and remained in Port Famine during the whole time we staid in that harbour. At first they lodged half a league to the northward, upon the sea-side; but, after some of our people had paid them visits, and presented them with some little articles, exchanging pieces of cloth, caps, &c. for their arms and ornaments, we so far gained their confidence, that they removed to some huts in the bottom of the harbour, of the same kind with many others which we had seen along the coast.

During our stay in Port Famine, the heavens were seldom free from clouds; the prevailing winds were from SW. to W. and often very fresh: on the 19th, in the morning, there were even some breezes from the N., but of short duration, nor did they extend out into the channel of the Strait, as we learned from our boat, which was at that time on her return from Tierra del Fuego.

On the morning of the 20th of January we sailed from Port Famine, with a clear sky and easy wind from WSW., for Cape St. Isidro. Soon after we got under-weigh it fell calm, and the vessel was driven back towards the shore by the current; but, by the help of our boats, we kept her at a proper distance until 5 P.M., when a little wind springing up from the SE. she soon got clear of the land. The wind afterwards kept changing from SE. to S. which forced us to resume our usual but painful employment of tacking; at night-fall, which had but an indifferent aspect, we discovered a patch of sea-weed, which always indicates a rocky bottom and shallow, or, at least, unequal depths of water, a good way out from the shore of Tierra del Fuego.

These circumstances proved how dangerous it was to continue under-sail during the night; we therefore resolved to steer for the Bay of Varcarcel, which we had before examined. In this project, however, we were disappointed; for it fell calm, and the vessel was carried by the current near to a ledge of rocks, so that we were obliged immediately to drop anchor in 15 fathoms. The bottom of the Strait is here so steep in some places, that, having dragged the anchor a little over, it came into 83 fathoms directly; but, getting out another, we kept the vessel steady.

Next morning we weighed for the above bay, (Varcarcel,) with very little wind from the SW. and, consequently, against us: at length, however, we got to the bay, and anchored in it in 40 fathoms, employing the remainder of the day in mooring the frigate, which we were never able to do to our satisfaction, notwithstanding the shelter it affords, on account of its narrowness and the great depth of water.

The natives followed us hither, having travelled along the shore eleven leagues, from the place where we first saw them. It appeared to us, that their only motive for following us, was their desire of increasing their wealth with the presents they frequently received from us; which, however trifling in our opinion, were to them of great value. That curiosity had no part in procuring to us their company, we were perfectly convinced, from repeated observations; for that passion which seems inherent in rational and in many brute creatures, did not appear to us ever to have entered the hearts of these men.

All the evening and night of the 22d, were employed in astronomical observations, to determine the position of Cape St. Isidro, which forms the east point of the Bay of Varcarcel; as also to ascertain the difference at that place between the magnetic and the true meridians.

On the morning of the 23d, with the wind at N. and NE., we set sail from this bay or harbour, but had scarcely reached the middle of the channel when it fell quite calm. In the evening there was a little wind from SW. to W., with which we tacked off and on, to make Cape Forward, which is the most southerly extremity of the north coast of the Strait; and is formed by a hill of moderate height, called by Sarmiento El Morro de St. Agueda.

Towards night, the weather being easy, we resolved to continue under-sail, taking care to keep a good look out for the shore on that side when we stood to the northward, because the water is so deep close to the land, that we might have ran on shore, before we had any suspicion of our danger.

The greatest part of the next day passed in calms; however, partly by light airs of wind, partly by towing, we at last got to Cape Forward, whose latitude we found, by a good observation at noon, to be 53° 55' south.

Although this position differs 10' or 12' from that laid down by some other navigators, yet, we trust, it is nearest to the truth; for, having determined the longitude of the same cape by Arnold's time-piece, No. 71, and the position of Port St. Joseph, observing its latitude with the quarter circle, the situation we assigned to Cape Forward, from a high mountain at that harbour, agreed exactly with the difference of latitude and longitude between the two points of observation deduced from the time-piece.

This day, whilst the frigate was under sail, we detached the boat to both sides of the Strait, to examine the harbour and roads; and, having passed the night in the same manner, about noon of the 25th we came to anchor in Port Solano, sufficiently tired with the voyage of these three days, which was only eleven or twelve leagues; with keeping under sail for two nights with some hazard, particularly in the second, when the wind, always contrary, blew with considerable violence, and with a continual succession of tacks, in all which we experienced the excellent properties of our vessel.'

In the Bay of Solano (called by the English, Wood's Bay) we remained four days, during which, the weather was so dark and cloudy, that we could make neither observation nor survey beyond what was necessary for laying down the draught of the port. The bay or port of Solano is defended from the winds from ESE. to N. and SW., but entirely open to those from the southward; and is so much narrowed by a bar formed by a river that falls into it, that a vessel at anchor has scarcely room to turn.

This disadvantage we learned from experience; for, during all the time we lay there, we were never a moment at our ease; and, even during some severe gales from the S. our stern came to be in 3½ fathoms water.

The harbour is sheltered on the W. by a high mountain,

forming a peninsula resembling that of Gibraltar; at the foot of it are to be seen a number of large fragments of rock, composed of petrified shells, which have fallen down a great while ago. The trees which have sprung up on the upper part of many of these fragments, may be considered as a most unequivocal proof of their antiquity, in a country where vegetation must make but very slow progress.

On the morning of the 29th, notwithstanding that the wind was at WNW we sailed from Port Solano, in search of a better situation, and made every exertion to arrive at the Bay of Cape Galan. The cloudy weather, showers of rain and snow, which had continued to succeed each other, at length disappeared; and we took advantage of the wind that prevailed, of moderate strength, to make several runs, by which means, at night, we had reached within two short leagues of the desired place of anchorage. Our intention now was to continue under sail all night, notwithstanding the dangerous navigation amongst the shoals of that part of the strait: however, having done so for a short time, at $9\frac{1}{2}$ P.M. we suddenly fell into nine fathoms sandy bottom, although we were still above a mile from the shore, we therefore came to anchor, and sent out the boat, to sound, and procure other information. The night, luckily, was very quiet; and, in the morning, we discovered that we had come to, on a bank that runs out from the Bay of Gaston, but which is not noticed in any former account of the strait. It may serve for anchorage, either in calm weather, or to avoid being under sail in the night time.

Early in the morning of the 30th, we observed coming out of the Bay of Gaston (called also by the Dutch, De Corde's Bay) seven canoes, containing, we reckoned, about 73 natives, men, women, and children. They approached us with loud cries and shouts, one of the canoes, advancing before the others, came within musket-shot, when the crew redoubled their noise, repeating frequently the word *pecheri*, holding up their arms, with feathers in their hands. We returned the most friendly answer we could, showing them some pieces of cloth, on which that canoe came alongside, and four of the stoutest of the crew immediately mounted the deck without the least fear. One of them, taking on himself the office of master of the ceremonies, introduced the other three, and addressed himself to one of our officers, making him understand that he wished to see our commander; and, being directed to him, again began his horrid vociferation, intimating, as we supposed, that he requested permission for the other canoes also to approach. In reply, we gave him some trifling present, on which he immediately made

Voyages and Travels, No. 5, Vol. II. E

signals to his companions, who soon arrived; and, in a short time, the ship was covered with Indians.

They resembled those we had seen in Port Famine, in language, dress, and manners, but seemed to show more activity and vigour in their actions than the others. They demonstrated the utmost indifference at the sight of our vessel, and of so many things entirely new to them; and, from what we observed at that time, as well as from what we learned during our residence in Port S. Joseph, we were more and more convinced of their miserable state of existence.

At $8\frac{1}{2}$ A.M. on the turn of the tide, which till then had set slowly to the SE. we set sail for Port S. Joseph. At 2 P.M. the natives left us, who had remained until that time on-board, perfectly quiet and contented; and, having understood our direction, they went on before us in their canoes.

From the time that we anchored on the N. of Point S. Maria, six leagues to the northward of Point S. Anna, at Port Famine, we had not perceived any considerable setting of the currents or tides; so that our way hitherto had been made good merely by force of tacking, and against the wind.

At three P.M. of the 30th we succeeded in anchoring in the bay or road of Cape Galan, that cape bearing S. 78° W. and about half a mile from the east shore of the bay; where we remained, without any remarkable occurrence, until the 2d February, employing the time in taking a draught of the bay, and in adding to our stock of wood and water. At night, some officers being on-shore with the launch making astronomical observations, the wind freshened so much, with dreadful squalls, that we expected every moment to be driven from our anchors; we therefore immediately let go a second, in the firm determination to hazard all we had on-board sooner than to abandon those who were on-shore; at the same time making ready to get under sail, in case, as we expected, our cables should give way; although, in that event, it would have been next to impossible to have avoided falling on the rocks close to leeward, on which we heard the waves dashing with fury, which, together with the darkness of the night, increased the horror of so cruel a situation. We can safely affirm, that the wind this night was the most violent that we met with in these parts; and that our cables must certainly have failed, had not the tide happened to set in a different direction from that of the wind, which came over the high mountains. This storm abated somewhat at day-break; but, in a few hours afterwards, returned with great fury, and so continued for two days together. This event having taught us how much we were exposed while lying in the outer bay, considering the state of our anchors and cables, we resolved to enter the inner harbour, called Port Galan or S. Joseph, which promised entire protection, being completely land-locked on all sides, with excellent bottom and convenient depth of water, from four to five fathoms mud. That of the bay is in general coral, with some spots of sand, of very small extent, as we discovered after minute investigation; consequently, very improper for cables such as ours, which had already been so severely treated.

The weather continued cloudy and foul all the remainder of this month, (February,) the sun being visible only for some short moments, and the wind constantly from WSW. to WNW. but not always equally strong.

From our entering the inner harbour until the 12th, the boat was employed in examining the neighbouring shores of the continent, in order to carry on our chart of the strait, when we had an opportunity of surveying to our satisfaction the Port of St. Miguel and Bay of Gaston.

From the earliest discovery of the Strait of Magellan, it has always been an object of research, to find out some communication with the South or Pacific Ocean; which, being free from such dangers and difficulties as those which abound in the western part of the strait, might still permit navigators to enjoy the benefit of the ports and refreshments furnished by its eastern part. M. Froger, upon the informations of M. Marcand, and since him, M. de Bougainville, have much recommended such a channel of communication; stating with great earnestness the information they had collected, which might be assisting in the enquiry. That we might co-operate in this business, we ascended a high mountain in the neighbourhood of Port S. Joseph, from whence we could discern a number of channels intersecting the Tierra del Fuego; and, being provided with all the notices handed down in the accounts of former travellers, we set out with the boat on this voyage of discovery, leaving directions with the officers, who remained behind, to take charge of the frigate: that if the wind should become favourable, they should leave Port S. Joseph, and remove to Port Candelaria or Tuesday Bay. Our want of anchors and cables was a very sufficient reason for not exposing that vessel, in such intricate and unknown passages as those we were about to examine; where either the opposition or the want of wind, joined to great depth of water, and multitudes of dangerous eddies, might bring her into such difficulties as we might never be able to overcome.

We set off, then, in the boat on the 13th February, and coasting along part of Tierra del Fuego, opposite to Port S. Joseph, entered the great bay of S. Simon. Our course was in general south, taking, as we passed, observations of bearings and distances, and never omitting, when we found a place where the boat could put in with safety, to ascend some hill or high ground, in order to acquire a more distinct notion of the different islands and channels which compose the archipelago or groupe called by the general name of Tierra del Fuego. Notwithstanding that the excessive depth of water, and the bad quality of the bottom, even close to these high precipitous shores, announced to us early, that it would be impossible to meet with a passage among them practicable for any vessel that did not make use of oars, we still continued all the evening in this labyrinth of islands, almost inaccessible on all parts, and of which it would have been equally difficult, as useless, to make an accurate survey. When night drew on, we turned to the westward, to find a proper place of shelter for our boat, since both the weather and want of light made it impracticable to continue our investigations, and where we might enjoy some repose. A cascade of water not far from us, an abundance of excellent shell-fish, a short stretch of beach, where we might draw up the boat and pitch our tents, all within a spacious bay, not destitute of wood, with the entrance defended by some small islands: these circumstances soon determined us there to fix our abode for the night. We accordingly landed, and put every thing in order; but, at half-past eleven at night, the tide, for which we had not made sufficient allowance in pitching our tent, came to interrupt our repose; so that we were obliged immediately to remove a little higher up, to a place where were the ruins of two huts, placed in that situation by the natives, well acquainted with the course of the tide.

Next morning we continued our researches in the boat; and, as the wind was northerly, we directed our course to the southward. The channels through which we passed, were, in general, narrow, and of vast depth, excepting one, in which there was no more than one and a half fathoms, and was besides wide and open. To this cluster of islands we gave the name of our commander, calling it the Labyrinth of Cordova.

From this station sundry channels presented themselves: in one of which, that appeared the most free from embarrassment, we proceeded for some hours, until we discovered another passage, which terminated in the Southern Ocean; and which, combining all the informations we had been able to collect, seemed to be that through which Marcant had sailed, giving it the name of his own ship, Channel of S. Barbara.

Having thus ascertained the communication outwardly, it as next of importance to discover that inwardly with the strait; for the channel we had followed, on entering this archipelago, was, beyond doubt, impracticable for any vessel of even moderate size. We therefore returned, and coasted along the island of S. Cayetano; and, on arriving at its S. and W. point, we found that what had appeared to us, from the mountains of Port S. Joseph, to be a deep bay, was, in fact, the true entrance of the Channel of S. Barbara; but that it was so crowded with islands, great and small, that no vessel could prudently attempt to go through it, especially considering the rapidity and contrariety of the currents which prevail throughout it. Thinking it useless to continue any longer our examinations of this part of Tierra del Fuego, we returned to Port Galan, where we found the frigate precisely as we had left her, the N. and NE. winds, which we had enjoyed in our excursion, not having extended to her.

We shall not be accused of rashness in affirming, after this enquiry, that there are many communications from the Strait of Magellan to the Southern Ocean; but that the navigation through any of them will never deserve to be undertaken or recommended, since the course of the winds, the currents, the narrowness of the channels, and, above all, the difficulty of anchorage in them, must always be insurmountable obstacles in the way of the navigator.

Agreeably to our purpose of surveying what remained unexplored of the Strait of Magellan, by means of our boats, without any more hazarding the frigate, which had thrice attempted to advance to the westward, but had been as often driven back by contrary and violent winds, running several times the greatest risk of utter destruction, and considering the frail state of our cables and anchors, and the unfavourable nature of the greater part of the bays and roads to the westward, we again sallied forth in the boats, in continuation of our labours, not intending to return to Port Galan until we had finished our remarks on the whole strait.

For this purpose two launches were fitted-out, and manned with several of the most experienced officers and seamen of the ships. Leaving Port Galan at 1 P.M. with a light air from the WNW. we took advantage of the current setting westward along the north shore; although, in the middle of the strait, it set to the eastward. On turning out of the bay, we came up with a canoe and a party of Indians, employed in cutting-off for food pieces of a whale, which they had fastened to the rocks. In the evening, we set up our tents on a spot of beach where were the ruins of some huts; and, to guard against a surprise, kept a regular watch, as on ship-board. Setting off next morning at 3 A.M. with a light breeze from SE. we entered the channel of San Geronimo; determined to examine it completely, to discover whether, as had been supposed, it communicated with the opening called Buckley's Channel, farther to the westward. That channel had never been examined by any navigator; and, to ascertain its nature and communications, seemed to be an object of importance in the geography of South America. The tide was now setting westward; we therefore made considerable way with our oars: but a heavy and incessant rain greatly incommoded us. On entering the channel of San Geronimo, we found it to extend NW. by N. having on the north coast, five miles within the entrance, a spacious bay running two miles into the land, and as much broad at the mouth. This is called in the English charts the Bay of Isles, from some small islands at the entrance. It is not however of importance; for no small vessel will probably ever stretch so far into the channel, which is there only one mile in breadth,—the current setting to the SE. with a rapidity of four miles an hour; and its contact with the counter current to the NW. along the north shore produces some very troublesome eddies. Along that shore are several other small bays, in all of which are streams of fresh water, poured down from the surrounding mountains, covered with snow in the midst of summer. In one of those bays we landed, to refresh the men, exhausted with many hours' hard labour at the oar. There also we measured a base for determining the chart of the channel. The south shore consists of lofty mountains, descending perpendicularly over the water, and forming a straight line of inaccessible precipice five leagues in extent from the entrance. The north shore, along which we ran, is low, near the water, green, and wooded; and the water near it is shallow, when compared with that on the south shore, where no bottom was found with a line of 40 fathoms. It was low water when we put into the above cove; but the current still set rapidly to the SE. in the mid-channel: high-water we found to happen at $12\frac{1}{2}$. Continuing our course up the channel, we found it enlarge considerably, forming a spacious bay, opening into which were two channels pointing the one NNE and the other W. In this spot the irregularity of the currents was great and sudden. At the entrance of this open bay are three islands, called in the English charts the Two Brothers, because one of the three is only a bare rock, while the other two are green and wooded. Proceeding still farther, we discovered that the channel on the W. led into another bay, six miles deep and five miles broad. From the north point of this last bay the land trends ends northward to the mouth of the channel, which extended to the NNE., in which bay are some small islands. The east coast of this channel is low, and verdant on the water, but the interior mountainous; the west coast is, on the contrary, a continued range of lofty perpendicular cliffs. As far as we could see up this NNE. channel, it appeared much broader than the external channel which opens into the strait, and to extend to a great distance in the same direction, agreeably to the course of the mountains. To follow that channel towards the interior of Patagonia would be, doubtless, very desirable; but, besides that such an attempt, in our circumstances, would have been very hazardous, it would have led us quite away from the peculiar object of our orders. We therefore turned towards the bay on the west, in which we imagined a passage might be found to Buckley Channel, or some other part of the strait. Our rowers being fatigued, at $6\frac{1}{2}$ P.M. we put in to the north shore of the bay; but, finding neither beach nor level where we could place our tents, were obliged to pass the night, some in the boats, and others on the rocks.

About three next morning, when our men were fast asleep, the tide flowed up to those on shore, and forced them to embark, drenched with the rain, which had poured down the whole night. Perceiving in the bottom of the bay a passage of no great breadth indeed, pointing WNW. we made towards it, and found that it opened into what may be called a lagoon, two leagues long and one mile broad. The water in it had no sensible motion, and was much fresher than that on the outside. The depth of water was considerable, excepting near the inner end, where it shallowed gradually, corresponding to the gentle slope of the adjoining land. In all other parts, the lake is enclosed by high steep mountains. The lake was covered with incredible numbers of penguins, of the same kind with those in all other parts of the strait. It was now evident that no communication existed, in that quarter, between the channel of San Geronimo and the proper Strait of Magellan; and that the other branch extending NNE. led away in a very different direction: when Sarmiento therefore affirmed such a communication to exist, he must have misunderstood the natives, from whom he had his information. Being satisfied on this point, we placed, in a projecting point, at the bottom of the lake, a bottle, containing an inscription, which noticed our observations on the spot.

Early next morning we departed with a fresh breeze at WNW. a smooth sea, and a rapid current to the SE. But, when we reached the W. end of the channel of San Geronimo, the wind came to blow so hard from the W. in addition to a current of three miles an hour, that, seeing every appearance of an approaching violent gale, we made away for the cove where we had dined the day before. There securing our boats, we pitched our tent on the only spot which, by its elevation, seemed to be beyond the reach of the tide. We lighted a fire, to dry and warm ourselves; for we all suffered severely from the incessant rain and cold, although on that spot Fahrenheit's thermometer indicated a temperature of 50 degrees. There, nevertheless, we slept till past midnight, when we were called up by the officer on watch, because the tide had risen and entered the tent. Hastening to safe ground, we passed, in the open air, the most inclement night that wind, rain, and cold, combined, could well produce. The spot we occupied was completely insulated by the rising tide, but we were safely carried off by the boats.

The violence and the variety of the currents in San Geronimo, are not less dangerous than surprising. In the middle, and along the northern shore, they set almost continually to the SE.; but, on the southern shore, they alternate with the tides. The meeting of these opposite streams at the mouth of the channel, is, at some periods, most dangerous for boats or small vessels; which, without a strong leading wind, will be unable to keep their course. To the violence of these currents must in a great measure be ascribed the contradictory statements given of those in the Strait by navigators. The storm still continuing, we erected our tent within a grove near the water; where, by cutting and burning the brush-wood, we cleared a spot for the purpose. We also availed ourselves of the position, to measure another base, and continue our survey. The ordinary rise of the tide appeared to be from six to seven feet, but the westerly winds then raised it to near nine feet. The difference of highwater in a day was about 45 minutes, and, at full and change, it flowed till near four o'clock.

Early next morning, with a light wind from the N. and an ebbing tide, we cleared the channel of San Geronimo, and turned westward into the strait itself, having to the south the islands of Charles III. and Ulloa. They are both vast masses of rocks piled together in disorder; the former lofty and precipitous over the water, having on it a few stunted trees, supported among the rocks by the almost incessant rains and melted snows: its extent from SE. by E. to NW. by W. is 14 miles. Ulloa's Island, of the same nature, contains besides some very elevated summits: its extent from ESE. to WNW. is about the same, and its distance from the former island two miles. Between these islands and Tierra del Fuego, is a passage called Whale Channel; but neither it, nor the islands, was it our business to examine. Running therefore along the north shore of the strait, we were obliged to take down our masts, for the wind came away right against us from WNW., and put into Langara's Bay, to rest and refresh our boats' company.

Scarcely had we left that place, when the wind again increased, and the rain, which had not ceased for three hours in all since we began our voyage, poured down in torrents; so that with great difficulty were our rowers able to gain another creek, only a mile to the westward of that we had just left. Next day the weather allowed us to keep sea for only three hours, and at five on the following morning we arrived at Cape Notches, a very remarkable object. It advances a good way into the strait, and is well named; for the south front is a lofty perpendicular face of smooth rock, having before it a number of detached pinnacles, separated by notches or clefts. From this point we observed Cape Monday, on the south shore of the strait, to bear N. 55° 20' W. Continuing our course for some miles, we passed some inlets, which had formerly been supposed to communicate with the channel of San Geronimo; a notion which our examination of that branch of the strait proved to be erroneous. In the course of the morning we perceived a fire on the south shore of the strait, the only indication of inhabitants which had been observed since we left our ship: and, indeed, the whole region seemed fit for amphibious animals alone, and not for human beings. At last we arrived in the Bay of Brown Beach (so called by Sarmiento), a beautiful harbour, defended from all winds of the SW. and NW. quarters. It is in breadth a mile, and runs in about three miles NNW. The surrounding land is mountainous and well wooded, terminating by gentle slopes at a sandy beach,—a probable indication of good anchor-ground. In front of the west point of the entrance is a remarkable triangular island, about one mile and a half in extent, separated from the point by a channel three cables wide. The English charts make no mention of this bay, nor does it seem to have been visited by any former navigators. The breadth of the strait here does not exceed two leagues: both coasts consist of lofty precipitous mountains, less elevated indeed than those inclosing Port St. Joseph, but still completely covered with snow, even in the midst of the summer of that quarter of the globe. Upon the whole, so dismal is the general appearance of that portion of the strait, that it seems to be the result of some sudden, powerful, and violent concussion of nature, and not of the gradual operation of natural causes for ages together. Towards noon we put into a small cove, half a mile to the westward of the bay just described, where we refreshed ourselves, after a hard morning's work. The air was clear, the wind light from NE., but the heat of the atmosphere was not only unacceptable but inconvenient; for the thermometer in the shade, at noon, gave 17½° of Réaumur. Rowing still along the north shore at 3½ P.M. being nearly opposite to Cape Monday, in Tierra del Fuego, we availed ourselves of the wind, fresh at E. to run across the strait for that cape. At 7 P.M. the wind fell, and heavy clouds springing up in the W. we ran into a channel three miles E. from the cape. The channel divided itself into two branches, of which we took that which pointed SSE. It was enclosed by lofty perpendicular mountains, and seemed to preserve a breadth of one cable and a quarter; but the waters being very still, we concluded it had no opening with the ocean to the southward. As it presented no beach or spot for our tent, we returned to the other branch, which led us into a harbour formed by a cluster of islands, where the water was without motion. The land was less elevated than in the other channel, but still we could discover neither beach nor level for our tents; we therefore had to pass the night in our boats. The wind changed to W. and a heavy rain, from which we were but poorly defended by our tilt, poured down the whole night. It will be sufficient warning to future navigators to state, that the ground of this channel is rocky, and very uneven.

Setting out at seven next morning, we found the wind in the strait blowing hard from WNW. with a heavy sea, both too powerful for open boats like ours; we therefore ran into a small bay, of which Cape Monday forms the W. point. Here we had fortunately an opportunity, at 9 A.M. of taking five altitudes of the sun, by which, according to Arnold's time-piece, we found Cape Monday to be situated in longitude 5° 30′ 25″ W. from the meridian of Cape Virgins: by other observations at noon, the latitude of Cape Monday appeared to be 53° 9′ 46″ S. In the evening, by dint of labour at the oars, we gained a little more than a mile beyond Cape Monday, and discovered a very valuable harbour, which, by its uniform circular figure, and the regularity of the amphitheatre of mountains by which it was enclosed, seemed to be the work of art rather than of nature. The mouth of the bay is divided by a small island into two passes, of which, that on the east is wide enough for boats only; but that on the west is 27 fathoms English in breadth: and, in the part next the island, the depth of water is between five and six fathoms, on sand. Within the harbour the depth is from eight to six fathoms, in all parts excepting along the shore, where it is only four fathoms rock. The entrance is one mile and a half W. from Cape Monday, bearing SW. from the W. point of the large island, lying in what is called Buckley Channel, on the north side of the strait. In taking the harbour, there is no danger but what is above water, or pointed out by sea-weed on the surface. This admirable spot is unnoticed by all former navigators; only Sarmiento mentions his having anchored in that quarter, in a place which he called Narrow Harbour. We examined and made a draught of the port, but found both wind and sea too violent to suffer us to continue our voyage; we therefore passed the night in a small cove on its west side. We observed the ebb and flow of the tide, but were surprised to see that, during both, the current on the outside continued still to the SE. In order to make the best use of our time, we measured on the shore a base of 96 fathoms, which was as much as the ground would allow; and, by angles measured from the extremities to objects at a distance, we obtained another base of 378 fathoms, by which we calculated the bearings and distances of several remote objects, to carry on the series of triangles necessary for our survey of the strait.

The gale still continuing, next morning we had, luckily, a short view of the sun, by which and our chronometer (Arnold's) we found the longitude of Cape Monday W. from Cape Virgins perfectly to coincide with that previously calculated. We at the same time observed the variation of the magnetic needle, by a method not generally known, but first practised by the eminent French astronomers Borda and Verdun. Two observers took at the same time, the one the sun's altitude, and the other his angular distance from a remarkable object on the north coast, of which the situation had been well ascertained. We then measured the angular elevation of that object above the horizon, and carefully marked its bearing by the compass: we next, by the complements of both altitudes and the distance, found the angle at the zenith formed by the verticals. By the sun's altitude and his declination, and the latitude, we found the azimuth, from which subtracting the first angle found, the remainder gave the true bearing of the object observed on the land. By this process, the variation of the needle came out to be 22° 43′ 56″ E. of N.

The wind fell about 2 P.M. and we proceeded on our voyage. Rowing about two miles, we landed on a small island, to take the bearings and intersections of several objects before observed. The wind suffered us only to gain a spacious bay to the east of Cape San Ildefonso, called by the English Cape Upright. At the east side of the mouth of the bay is a triangular island, one mile and a half in length NE. by N. and SW. by S.; having between it and the main of Tierra del Fuego a channel one mile and a half broad, and three miles in depth, to the bottom of the bay. The bay is in breadth two leagues, and in the bottom, towards the east side, is a channel, running up S. by E. among the mountains. Between the island and the east coast we thought there might be good anchor-ground, but we found no bottom at 30 fathoms. The west point of this bay is low, and beyond it is another bay, less broad, but as deep as the former, being bounded on the W. by Cape Upright: for this reason Captain Wallis called it Upright Bay. He anchored on the west side of the island just mentioned; but that position is by no means to be recommended, for the ship, on account of the great depth of water, must be very close to the island, which will then be under her lee in NW. winds, which are the most prevalent in that quarter: nor can a cable be made fast to the trees on the island, which are all too small and weak for that purpose. The wind still freshening from the NW., we resolved to run under the east side of Cape Upright for the night, which we passed in our boats, not finding a spot of beach or ground for our tents.

About E. from the triangular island before noticed, appears, on the northern continent, the mouth of Buckley Channel, formerly supposed to communicate with the channel of San Geronimo; which communication we had already found not to exist. The breadth of the entrance, measured obliquely from N. to S. seemed nearly four miles. In it are two islands, of which the largest is high, precipitous, and hilly, destitute of all vegetation, being in length $2\frac{1}{4}$ miles. The channel points E. by N., and from the NW. point the coast of the continent extends without any noticeable object, gradually falling off from the south coast of the strait on to the projecting point called Cape Providence. The breadth of the Strait of Magellan to the east of Buckley Channel, is five miles; from Cape Monday to the island in that channel, it is four miles; but, to the west of the channel, the breadth is not less than $2\frac{1}{2}$ leagues. The height of the lands on both sides of the strait, is less than in the parts towards the middle of the strait; and nothing is to be seen but naked rocks, which present a most melancholy prospect: nor among the mountains are any observed of remarkable elevation above the others; all are, nevertheless, crowned with snow, even in the midst of summer.

In the morning after our arrival under Cape Upright, the weather being favourable, we found the magnetic variation $22°\ 39'\ 40''$ E. of N.

Taking to our oars at 7 A.M. the weather becoming easier, we soon arrived at Cape Upright, which is situated NW. by W. $4\frac{1}{2}$ leagues from Cape Monday. The cape runs a great way out into the strait, and is composed of a chain of lofty rocks, resembling on the summit a range of waves of the sea. The north face runs E. and W. two-thirds of a mile, lofty and perpendicular over the water, having, one mile E. by S. from it, a large rock above water, surrounded by shoals and sunk rocks, indicated by the sea-weed on the surface. On the north coast of the strait, N. eight miles from Cape Upright, is Cape Providence, before noticed; a remarkable head-land, consisting of two eminences, and having on each side a spacious open bay. To the west of Cape Upright the coast forms a deep bay to the south, having in the east part a number of islands and rocks under water: on the shore were seen some deserted huts of the natives. In the bottom of the bay is an inlet, pointing SSE. This is the Bay of Islands of Captain Wallis. Adjoining is another open bay, bounded on the west by the low flat point of Echenique, bearing WNW. ten miles from Cape Upright. The wind and current were now strong from the NW. and our people fatigued; we therefore put into a creek on the east side of a point in the middle of the bay, where we passed the night on-board. In making for that place, we observed soms Indians, who called to us; and scarcely had we secured our boats by the shore, when a canoe, with eight men on-board, came up to us. As they approached, they showed some symptoms of fear, and made signs to one of our officers to land, which he did, and then one of the natives did the same. This man seemed to be a chief, for he wore a lofty white sugar-loaf cap, made of a duck's-skin and feathers, by which he was distinguished from his companions. He was a stout young man, well made, but short; and, when he came up to our officer, he pronounced, with a voice and tone in which terror and courage were intermixed, a long discourse, of which we were unable to make out a word. Having received from us a few trifles, which he understood to be marks of our good-will, he called to the other natives, who all landed; and, having also received presents, they staid by us till night-fall. Their language we could perceive to resemble that of other natives of the eastern parts of the strait; and, although better shaped, they were of similar features and stature: their dress was also the same. They were delighted with our presents of beads, ribbons, and looking-glasses, which seemed to affect them as a child in Europe is pleased with a new plaything.

The wind continually increased, and the rain poured down with such force, that our provisions, as well as ourselves, were drenched, in spite of our tilts and awnings: nor was there a spot on which we could land for shelter from the tempest. On the following morning, the Indians returned, with a present of ducks, which we remarked to be all wounded in the head,—a proof of the dexterity with which the natives managed their bow and arrows in hunting. They gave us also three arrows, and strings of shells for necklaces and bracelets; and all without seeming to expect any thing in return. The construction of their canoes displays a superiority over the workmanship of the natives of the eastern parts of the strait. They are not made of frail bark, and badly put together, like the others, but of planks joined together by a strong cord half an inch thick, and the seams covered with a compound of leaves of some plant and a very adhesive clay, which prevents the entrance of the water. Each side of the canoe consists of two planks, cut away so as to diminish regularly at stem and stern. The bottom is a strong plank, broad in the middle, but sloped off to each end, and joined to the side planks in the same way as themselves. The sides are kept asunder by thwarts for rowing. These canoes are not well adapted for swiftness, but they are strong and steady, and not apt to sink by the admission of the water. The natives row as we do in Europe, and their oars are of a proper size; showing the natives to know the advantage of a long oar over the short paddles of the other Indians. They however employ a paddle to steer with, particularly when the sea is rough. All these improvements in their canoes show with what skill men, in all situations, adapt their contrivances to their needs; for, in the heavy seas and gales of the western parts of the Strait of Magellan, the feeble barks of the eastern and narrow parts would be wholly useless.

Having proceeded with one of our boats to survey the neighbourhood of our station, we discovered, in the bottom of the bay, the entrance of a harbour, nearly two cables in width. We sounded over it, but found the ground, in general, rock, with a little loose mud, and the depth to vary from $4\frac{1}{2}$ to 17 fathoms. On each side of the harbour the mountains spring up very high and steep from the water's edge, near which are a few stunted shrubs: its greatest extent is about half a mile. The entrance lies S. 42° 30′ W. from Cape Providence. Within the harbour is abundance of fresh water, from the melting snows on the mountains, and a little fire-wood may be procured: but we would not advise any vessel to enter it, on account of the bad nature of the ground. We nevertheless made a plan of the harbour, which the commander named, after one of our officers, Port Uriarte. In the evening, the wind abated, and, proceeding to a low spot on the E. of Point Echenique, we then took the bearing of several objects, in particular of Cape Tamar, on the north coast, distant four leagues. That cape is a broad promontory of lofty rocks, advancing into the strait farther than Cape Providence, and having on the summit two slender peaks. The width of the strait, right across from Cape Providence, is $3\frac{1}{2}$ leagues: the north coast forms several spacious bays; but the south coast is a continued succession of small curves, separated by projecting points of rocky cliff, trending more towards the W. from Cape Upright than it did to the eastward of that cape. Having been detained three long days in the miserable corner already described, we at last ventured out to sea; the weather becoming moderate, although the wind was at NW. We passed several deep inlets, separated by islands and points; but the wind being contrary, and the sea very heavy, we put into a small bay, which we supposed to be that of Sa. Monica, so called by Sarmiento. No circumstance of his description, however, corresponded to that bay, which was surrounded by high land, and no bottom was found at 45 fathoms. From these and other circumstances we were convinced, that some error exists in the account of his voyage through the strait, in the distance between the bay of Sa. Monica and Cape Upright. In that most unfit situation for our small vessel, we were overtaken by a most furious tempest of squalls and rain, particularly during the night, which, by the roaring of the winds among the mountains, and the dashing of the waves against the steep cliffs around us, was the most distressing of all our voyage. In the course of the two following days, we heard a dull sort of noise, of short duration, which we supposed to be thunder; but we were afterwards inclined to ascribe the noise to some explosion among the mountains, in which we believed volcanos existed, although not in visible conflagration. To this conclusion we were led, by having found, on the summit of one which we ascended, masses of scoria, or metallic cinders, perfected similar to those produced in a smith's forge. On the last day of our detention in that wretched spot happened the new moon, and it was high water at 5 minutes past 12 noon. The weather becoming more moderate in the afternoon, we at last proceeded on our way, and soon came to another harbour, small, but tolerably commodious; for the ground is good, and the bay is sheltered from the SW. and NW. quarters. The only defect is, that the surrounding land is low; but, being well wooded on those quarters, vessels may lie in the bay very safely, especially as they may be made fast to the trees, if necessary. The harbour runs in to the west about half a mile; and the wood and verdure around it, which are very uncommon in the western part of the strait, together with the runs of fresh water from the snows of the interior, make this harbour an object of value. From these and other circumstances, it seems highly probable that this is, in fact, the harbour of Sa. Monica: its distance is 14 miles about WNW. from Cape Upright, and SSW. from Cape Tamar. In this port we passed the night, having found there, as in all the other parts of the strait, abundance of mussels;

and, early next morning, the wind being easy from the NE. although the sea still set from NW. we put off on our voyage. As soon as we entered the strait, we had the satisfaction, at last, to behold Cape Pillar, the western limit of the strait, and of our dreary expedition. We marked also the position of Cape Tamar and Sa. Anne's Island, lying in the bay on its east side; as also numerous other islands on the north coast towards the mouth of the strait.

The motion of the boats was rather unfavourable to our observations; but we were able to fix Cape Pillar N. 62° 20′ W. from Cape Upright, or S. Ildenfonso. We then made for a point of land, to determine more accurately a number of positions for completing the chart of the strait; but the rain and thick weather soon hid them from us. Leaving that point, we came to a bay, in which are two harbours of good appearance, but in neither could we find ground with 45 fathoms, excepting close to the shore, where it was full of large stones, and of very unequal depth. A little farther, we came to two other harbours, of which the one runs in W. by N. and the other SE. The former is in length above three cables by nearly two in breadth, having in it a beautiful cascade, remarkable for the body of water and the noise of the fall. This harbour, afterwards named from another officer, Port Churruca, lies SW. by S. from Cape Tamar. The entrance is broad and clean, even close to the shores: few navigators will, however, be tempted to resort to it, except from necessity, when they can make fast a cable to the rocks.

Having landed on the west point, to make farther observations, we continued our course; but the wind soon again grew strong from the NW. and the sea much more formidable, in proportion as we approached the entrance of the strait. The south coast was all along lined with rocks and islands, some stretching out one mile into the strait. Notwithstanding the dangers of such a coast, the wind and the waves from the Pacific Ocean rebounding from the north coast, forced us to take shelter on a small extent of beach in a corner defended from all winds, where we remained all night. The weather was clear, which was no small relief to us, who for so many days had never seen the sun, and whose clothes and provisions were drenched with almost incessant rains. On the following morning, the wind had fallen, the sea was much abated, and the ebb-tide in our favour, we therefore again put to sea.

Rowing westward, we passed a long succession of rocky islands, till we came to a head-land, remarkable by its form and situation, although it does not project much beyond the general line of the coast. Its termination is bare and high, having the appearance of a hill sloping down to the north, and as if detached from the land by some violent concussion; around it are seen enormous masses, of the same nature, probably separated by the same cause. A distinguishing feature of this head-land, is a large black spot on the north face, which appears, when nearer, to be a vast deep cavern, within which the sea breaks with tremendous noise. On the west of this cavern are others on the level of the sea, but smaller; on which account, we called the head Cape Caves. It is situated S. by W. from the island on the north side of the strait, called, from its general appearance, Westminster Hall, and $30\frac{1}{2}$ miles westward from Cape Upright: Cape Tamar bears N. 86° 30′ E. About one league N. 43° 20′ W. from Cape Caves, is another, much more projecting; but of that cape, and the intermediate bays or harbours, we shall speak more fully afterwards: for we had, at this time, a singular combination of circumstances favourable to our pushing on to the end of our voyage. The air was perfectly calm, the heavens bright, and the water smooth; such at least it might be called: for, although the great swell from the Pacific still rolled into the strait, it gave us no extraordinary trouble. So favourable a state of the weather is not, perhaps, to be found in that region in the course of many months; we hastened therefore to avail ourselves of it, to terminate our series of triangles for laying down our chart of the strait, and, if possible, to take a meridian altitude of the sun at Cape Pillar. The cape already mentioned, one league to the NW. of Cape Caves, is noticeable by its elevation and projection, and by its north face, which appears as if cut down perpendicularly with an instrument: from this circumstance we called it Cape Cut-down. About $1\frac{1}{4}$ league westward from that cape, is the bay or harbour of Misericordia, (Mercy,) so called by Sarmiento, being the first place he entered after he came into the strait from the ocean: but more of this harbour afterwards.

On the west side of Cape Tamar the strait suddenly widens considerably, by the falling-off of the north coast, which disappears from the eye on the water. That cape forms the east side of a spacious mouth of a channel, running NE. into the continent; and, on the west side of that mouth, the strait is somewhat more than five leagues in breadth, which gradually increases, until, opposite to Cape Pillar, it is upwards of six leagues. All along the north coast appear a succession of openings, or channels; but, in the interior of these openings, we thought we could perceive land. In that case, the spaces which seemed to divide the openings are probably islands, stretching along before the coast, which is low, and was very hazy when we observed it: no dependance ought therefore to be placed on our account of it. To the westward of the island called Westminster Hall, are extensive groupes of islands, of which the southernmost we laid down by accurate observation: they were afterwards named, by our commander, Cevallos's Islands.

The coast of the continent, on the west of Cape Tamar, seemed to consist of bare precipices, destitute of vegetation; and, although both coasts of the strait are much lower than in the middle quarters, still the north coast is much less elevated than that on the south. Having the opportunity of a clear sky, when off Cape Caves, by means of 5 altitudes of the sun about $8\frac{1}{2}$ A.M. and Arnold's chronometer, we found that point to be in longitude 6° 45′ 32″ W. from Cape Virgins, and 1° 15′ 7″ W. from Cape Monday. The latitude, at the same point, by an observation at noon, was 52° 47′ 21″ S.

We cannot pass over in silence the singular occurrences of that day, which was the second day in which, during our voyage hitherto, the sun shone bright for some hours together. This may appear to many persons a matter of no great importance; but the effect of it on our minds is not to be described. For eighteen days the rain had never ceased to pour upon us, generally with great violence; we slept constantly wet, at one time in our confined boats, at another on the beach or the rocks, almost in the open air: our voyage had been greatly retarded by contrary winds, and the weather seldom allowed us to make the necessary observations for our purpose. We had besides been reduced to short allowance, by the injury done to our provisions by the rain. Any one of these misfortunes would have been enough to dishearten us; but all uniting together, were sufficient to destroy the hardiest constitution, and to overcome the most determined perseverance. But that one day, serene and temperate, which carried us to the end of our course, and enabled us to dry our clothes and stores, gave new elasticity to our minds, by shewing that we had vanquished obstacles which, had they been foreseen, would have discouraged us from engaging in the enterprise. At 3 P.M. having rowed twenty miles in eleven hours, we arrived at last at the point of Cape Pillar, the western extremity of the Strait of Magellan, on the coast of Tierra del Fuego; and that fortunate event we celebrated by hoisting the Spanish flag, and saluting it with seven times " God save the King." The eagerness and exertions of our seamen to reach the cape, which incessant tempests had made us almost despair of ever attaining, are not to be described; nor was their joy less, on accomplishing their dreary task; so much so, that it was some time before they could be prevailed on to take their necessary refreshment.

From Cape Cut-down to Cape Pillar, the south shore is so clean, that no island, rock, or shoal, extends half a mile from the land; it may therefore be run down without fear. The north side, on the contrary, which Narborough very justly termed the desolation of the south, seems to be only the fragments of a ruined world torn to pieces by violent earthquakes; for, out before it lie a multitude of islands and rocks, extending far from the shore; so that no vessel ought, on any account, to approach that side of the strait. It is at the same time to be remembered, that no one ought to run very close to the south shore, unless the wind be fair and steady; for the heavy sea from the Pacific from the NW. might, in such a case, be very dangerous.

The famous Cape Pillar, remarkable for its position on the south side of the west entrance into the Strait of Magellan, and for its elevation over the water, is still more so, for two peaks which rise on its summit, both inclining a little to the northwest. That on the east, which is the highest, is connected with a hill from which the cape itself projects; the peak on the west rises up like a great tower from a base on the edge of the water, on the west of the cape: from the general resemblance of that mass of rock to a rude pillar, when seen from the west, the cape probably received its name. The union of the strait with the great Southern Ocean, on the west of America, seems to be bounded by rocks of the same nature; for the quality and distribution of the strata appeared, as far as we could compare them, to be perfectly similar; and their separation has evidently been, originally, the effect of some violent convulsion; although it be also evident, that the sea has made, and is still making, great inroads on both sides of the strait: for the shores consist of high precipitous cliffs, before which lie, at different distances from the land, multitudes of fragments, and even rocky islands, of precisely the same appearance with the present shores. That portion of Cape Pillar which is washed by the waters of the strait, appears like the rounded summit of a low hill; but the opposite portion exposed to the ocean, is excavated in many places by the action of the waves in the solid rock. At the cape, the region called Tierra del Fuego bends round to the SSW. at least as far as we could observe from the boats. Half a mile in that direction from the cape, and close under the land, are two small islands, which we named the Spanish Long-boats; but we could not discover the cluster of rocky islands called, from their number, the Twelve Apostles, which must therefore lie farther out than is commonly supposed. By observation, we calculated the height of the western peak over the sea, to be 395 English yards, and the eastern 463 yards. While we were at the cape the weather was perfectly calm; but the swell from the north-west was such, that it was absolutely impossible to land in any part of the point, or even in the neighbourhood, so powerful was the back-draught from the perpendicular cliffs over the water. In the evening the weather grew hazy; so that, on the north shore, we could perceive only Cape Victoria, the western extremity of the strait, which we ascertained to bear N. 7° 20′ W. and S. 7° 20′ E. with Cape Pillar. Cape Victoria appeared to us as a broad even point, of no great height; but, as we could see it distinctly from the water, although at the distance of $7\frac{1}{2}$ leagues, its real elevation must be considerable: but we judged of it, by comparing its height with that of two round hills to the eastward, from which the cape seemed to proceed.

Having now arrived at the western limit of our expedition, we prepared to avail ourselves of the calm, to return to the eastward; for, as the coast from Cape Pillar offers no inlet or bend to afford shelter, in case of necessity, on to Port Misericordia, a distance of three miles, a change of weather, particularly in the night-time, would probably have been fatal. A wind from E. would have carried us out into the Pacific—from W. or N. would have forced us to leeward on the rocks; and the wind from the S. or SW. would have sent us over towards the north shore of the strait, six leagues off, but only known to be full of danger. The favourable calm, however, continued; and, when we had rowed $2\frac{1}{2}$ miles, we landed on the largest of the islands at the west point of Misericordia, hoping to have a view of the Four Evangelists, small islands in the middle of the entrance of the strait. Although very distant, and the atmosphere was cloudy, we were luckily able to discover them, and to mark their position, and that of several other objects of importance in our chart. Having finished our observations on the island, we deposited on it a bottle containing a paper, denoting the period and the object of our voyage, with the names of the officers in the two boats. This Island of the Bottle, as we named it, is of considerable elevation, consisting of layers of stone parallel among themselves, but all inclined towards the strait, manifesting their having been torn off from the adjoining precipices on the land. On the highest part of the island we found a vast quantity of sea-shells lying loose. They appeared to be all recent, and all of one kind, that is, of a species of mussel frequent along the coast. We could account for their assemblage in that spot only by supposing them to have been carried thither by the sea-fowl. Leaving the island, we rowed into Port Misericordia, which we found to consist of a bay and two small creeks. The points of the entrance lie ESE. and WNW. two-thirds of a mile asunder, and the port runs in SW. one-third of a mile; it then contracts, and forms a small harbour $2\frac{1}{4}$ cables long to the W. and about as broad in the middle, which terminates in another, resembling a wet-dock, one cable long, and three-quarters of a cable in its greatest breadth. In the outer bay is good anchorage towards the west shore, in from nine to eighteen fathoms sand, excepting where sea-weed is seen, which always indicates rock and stone, and generally shallow water. In the inner parts the ground is the same, but the depth of water is 18 fathoms, close to the shore. The port is excellent in its situation, for it is sheltered from all prevailing winds and the heavy sea from the W.: a number of vessels may be acommodated in it, and the holding-ground is good. Being only three miles within Cape Pillar, it may be of great service to ships driven into the strait by the winds and currents from the Pacific, as it furnishes fresh water in abundance, plenty of shell-fish, and fire-wood sufficient for a long voyage. The entrance is clean and easy, there being no danger but what is above water; and the bay is easily known by its distance from Cape Pillar, and by the islands before its west point. By the time we had finished our remarks on the harbour, we were overtaken by the night, which we passed on-board, as we found no spot to pitch our tents. It rained the whole night, and in the morning the NW. wind set in, when we ran to the E. to survey the harbour on the east side of Cape Cut-down, or Tuesday, of Captain Carteret. We found it to afford an excellent station in westerly winds, for its points lie NNW. and SSE. and it is enclosed on all sides by high land. The opening is half a mile wide, and the harbour runs in W. to a narrow pass, two cables over, when it turns WNW. widening and lengthening for one mile and three-quarters. In the outer harbour the north shore affords both shelter and good ground, there, as well as in the middle, consisting of fine sand, and, in parts, gravel: the depth of water from 12 to 22 fathoms. Farther in are several rocky shoals, but they are always known by the sea-weed on them. In the middle of the great inner harbour the ground is generally rocky, and the depth very unequal, from 22 to 42 fathoms: it will therefore be most advisable to come-to in the outer harbour, under the north shore, for the wind is apt to change there from WNW. and W. to NNW. and N. probably from the reflection from the north coast of the strait. The most dangerous wind is that from the E. which is indeed not frequent: in such a case, a vessel ought instantly to get under sail. In the harbour, wood and water may be procured in abundance; and, being only 8½ miles from Cape Pillar, and easily known by Cape Cut-down on its west side, as also having a clean easy entrance, Tuesday Harbour cannot fail to be of singular service to ships driven in from the westward. The east point of this harbour is the west point of another spacious bay, one mile and a half between the points bearing NW. and SE., and half a mile deep, forming nearly a semicircle. In the bottom, the bay contracts to half a mile, by points advancing on each side, and situated N. by W. and S. by E. There begins a well-defended haven, which runs into the land WSW. one mile and a quarter, and terminating in two small coves. The bay contains very deep water; and it is only on the west side, near the land, that proper soundings are found,—from seven to twelve fathoms, sand and coral. Near the mouth of the inner harbour the depth is greater, and the ground rock. In the inside are some shoals, shown by the sea-weed; but on none of them did we find less than six fathoms. Fresh-water, wood, and shell-fish, are not wanting in this bay; but few vessels will, nevertheless, enter it, when they can take Tuesday Bay so near it, and nearer the mouth of the strait. The outer bay was called Truxillo Bay, and the inner Port Rivero.

We had now completed our survey of the coast of Tierra del Fuego, all the way to the westward from Cape Monday to Cape Pillar, according to our instructions; and we feel ourselves warranted to assure all future navigators, that they may, without apprehension, follow our remarks and counsels. Of the north coast of the strait we had determined the situation of the most remarkable points, which seemed to be all that was requisite in a tract particularly exposed to the prevalent winds and currents in that part of the strait; and which presents only a long succession of islands and rocks, which no navigator, excepting in very peculiar circumstances, will ever approach. It was, besides, specially enjoined us to confine our enquiries to the southern coast alone.

Our commission being now terminated, we resolved to take advantage of a light breeze from the NW. to return to our ship; but, no sooner had we left the bay, than we met with a heavy swell from the same quarter, accompanied by a quick succession of calms and squalls. These sudden changes we ascribed to our being near the land; we therefore made off for the middle of the strait, in the hope of having the wind more steady and moderate. Steady indeed we found it to be, but so strong, that it was with the utmost difficulty we could prevent our small open boats from oversetting. In this desperate situation, it being impossible to run under either shore, we could only strive to keep our boats before the wind, and so were driven for above ten miles; and at last, near four P.M., managed to get into an open bay on the south coast, which we justly named the Bay of Good Fortune, lying to the WSW. of Cape Tamar. Our provisions were now greatly reduced, and our voyage had occupied much more time than had been supposed to be necessary; the tempest had, however, scarcely abated, we therefore resolved to remain all night in that fortunate station. Next morning, notwithstanding that the sea and the wind continued as boisterous as before, with the addition of incessant rain, we left our birth, and, in order to clear the rocks and shoals, stood out a couple of miles from the land. Keeping our boats' heads as much as possible before the wind and sea, we run down above 37 miles, and at last got under the lee on the east side of Cape Monday, where we again observed several bearings of objects on the opposite coast. Next morning the gale had abated, but the swell continued; we nevertheless managed to cross over to the north shore, near Cape Quade, where we put in for the night, and on the following day had the unexpected happiness to arrive in Port Galan, or St. Joseph, where our ship still lay, and where our return began to be despaired of; our expedition having lasted no less than 22 days. Notwithstanding the duration and hardships of our voyage, especially for men accustomed to the very opposite climate of the south of Spain, only one of our company was taken ill. This was a seaman, who, soon after we left the ship, was attacked by an ague, but who perfectly recovered his health, in the midst of our toils, without the use of any remedy.

Whilst we lay in Port Galan or St. Joseph, we had considerable intercourse with the natives, whom we called the Indians of Tierra del Fuego; both because that tract of country seemed to be their usual place of abode, and that Captain Cook found the same nation established on the most southerly parts of it. We never discovered in them any depravity of inclination; not even that which might be deemed so natural to men in their situation, of making themselves masters of any thing belonging to us, which could only be done by theft: but, far from attributing this conduct to any sense of moral virtue inherent in these men, we were rather inclined, after much attentive observation, to account for it on the ground, that nothing is capable of moving the torpid indolence of their hearts.

They often left us; at one time altogether, at another by parties: at last, on the 24th February, they entirely forsook us, on account, as we supposed, of the death of one of their boys; for it is their practice to abandon those places where any unfortunate accident has befallen them. Our chaplains privately administered baptism to this child, who, as the surgeons thought, was under two years of age, giving him the names Antonio Joseph Julian. We buried him in the point of low land forming the west side of the entrance of the harbour; and, in the interior of the wood, near the same place, we deposited the body of a seaman, who, in consequence of an ague with which he was attacked before we left Spain, died on the last day of this month.

In one of our excursions to the mountains which surround Port Galan, we found a bottle sealed up, containing a long Latin inscription, and placed there by M. de Bougainville, as he went this way, on his voyage round the world in 1768. In imitation of his example, we left another of the same kind; and gave to the mountain on which these monuments were left, the name of Cerro de la Cruz (Mountain of the Cross).

Winter already began (end of February) in this inhospitable climate to give notice of its approach: our cables suffered greatly from the hurricanes from WNW. to SW. which blew with but little intermission; so that we were much alarmed for them; which induced us speedily to come to some resolution respecting our future motions. A council of officers was held, in which it was resolved, that, in this case, our best way to conform to the orders of his Majesty, was directly to return to Europe, without exposing, by an ill-timed and unnecessary temerity, the vessel and the people to fresh disasters, and the public treasury to additional charges.

The ship's company, although greatly fatigued with the constant and severe duty of the frigate in these tempestuous regions, and with the various excursions in the boats, during which they had but little relaxation from their toils, by night or by day, still continued to enjoy good health, and was therefore no inducement for us to embrace this resolution, notwithstanding that we were in want of many necessary articles, particularly for preserving the health of the men; and that we were under the necessity of diminishing, by one-fourth part, the daily allowance of provisions, in order to ensure a sufficiency for our consumption.

We had now only to wait until the wind should become favourable to put our design in execution; and, in the mean time, set about examining the motions of our time-pieces,— an operation as easy in itself as irksome in this climate, when the heavens seem to be the declared enemies of astronomy,—it being no uncommon thing, in this gloomy and horrid region, even in the summer, to pass 15 or 20 days together, without a sight of sun, moon, or stars.

At last, on the 11th of March, after a residence of thirty-nine days in Port Galan or St. Joseph, with winds so constantly unfavourable, that it would have been impossible for us to have proceeded farther to the westward of the strait, we set sail, on our return to the eastward, with a fresh gale from SW. and, at noon, ratified our observations of the latitude of Cape Forward, by which we confirmed other positions which had been determined relatively to that point. The wind abated much in the night; however, we kept under way till next day, when we had it from NNW. to NW. which obliged us to return to our usual exercise of tacking. It fell almost calm in the morning, when we came to anchor about three leagues N. from Point S. Maria, and half a mile from the land. The night continued calm, with dark cloudy weather.

Soon after day-break we weighed, having a fine wind, but unsteady from the W.: by it, however, we had the good fortune to clear, without difficulty, the very dangerous pass between Elizabeth Island and the Isles of S. Martha and Magdalen, on the E. At 1 P.M., favoured more by the current than by the wind, we ran through the narrow pass of S. Simon; and at half-past four came to anchor in the Bay of S. Gregorio, having the cape of that name S. 36° W. distant three or four miles.

Observing the beach to be lined with Patagonians, in great numbers, all on horseback, and attended by multitudes of dogs, some of our people went on-shore, and were received in their wonted kind and friendly manner by these natives, who happened to be the same we had seen on our passage to the westward. They brought seven of them on-board, who eat heartily of what we set before them, showed much inclination to smoke tobacco, and a vehement desire to obtain swords or hangers, of which we exchanged a few with them for the skins of lamas and zorillos. Some women had come down to the shore with the men, but none of them chose to come off to the frigate.

On the 14th we again set sail; but, by a singular contrariety of the winds, we found them now as much to oppose our going out of the strait as they had formerly resisted our coming into it. They began to blow from the NE. quarter, which forced us to tack; and, in standing over to the southward, we fell quite unexpectedly into shallow water. As it was then 5 P.M. and we saw ourselves surrounded almost with shoals and sand-banks, we set about finding some good anchor-ground, which we met with about two leagues S. 66° W. from the mouth of the narrow pass of Esperanza, otherwise called the First Pass.

The wind, which had freshened in the fore part of the former night, this day increased much; and it being impossible to set sail, as the tide was strong against us, our situation, for the greatest part of the day, became extremely critical. The fate of the frigate depended on the resistance of cables, already severely handled, opposed to a furious wind, a heavy sea, and to the current of the tide, which, at 8 P.M. would take the same direction with the wind. If our cable should give way, we would remain in a narrow pass, between two mouths, engaged in a small space, where it was impossible to tack in the dark, or indeed to take any other course for our safety, on account of the shoals running out a long way from both shores, as well as of the incalculable alteration in the situation of the ship, produced by the current. To let go another anchor was useless, as its cable, being single and less strong than that one already out, it would certainly give way when the first should fail. But, most fortunately, about 10 P.M. the wind abated, and we began to conceive good hopes of at last being delivered from such terrible situations. In the evening of the 16th, the wind being northerly, but very feeble, we began to proceed out of the strait: however, it soon changed to the NE. which obliged us to turn up against it,—an employment disagreeable enough in itself in these parts, but particularly so now; as we could perceive that this cross wind did not extend much behind us to the westward. This opposition in the direction of the winds happening several times before we got out of the strait, showed that those from the NE. however strong they may be at the entrance, never extend far into that part, where the westerly winds seemed to have established their unrivalled dominion.

The tide was now setting towards the NE.: we ran through the first pass, and, at 7 P.M. we came into anchor in 7 fathoms sand and stones, a little to the eastward of the mouth, and very close to the shore of the continent on the north. Soon after this, the tide changed its direction to the SW. and the water rose to 9 fathoms; but, continuing still to run in that direction, at 12 it began to diminish, and, at 5, remained under $3\frac{1}{2}$ faths.

It will not be difficult to conceive our distress, on finding ourselves, in addition to all other unfavourable circumstances, now in the midst of an unknown current, of whose duration we were totally ignorant. However, notwithstanding there was a thick fog, which prevented us from seeing the land, in a situation full of dangers, which by land-marks alone can be avoided, we set sail, and, by the help of the tide, and variable light winds, at last escaped from these perils. We conducted ourselves now wholly by the soundings: at first the depth increased, then it diminished, and we came on a shoal of only $3\frac{1}{2}$ fathoms: we then changed our course, involved in endless doubts, which were rather augmented than removed by the line; as it at one time pointed out as favourable a direction which, at another, led us into new dangers.

Easy winds and the tide had hitherto preserved us; but, as it would have been highly improper any longer to commit to such conductors the fate of the frigate, and seeing the fog growing every moment more dense, deprived of the sight of even the nearest objects, we dropped anchor in $8\frac{1}{2}$ fathoms, when we waited for a better conjunction to continue our course, although we were totally ignorant of our position; so that, in case of any accident, such as our cables giving way, we were as likely to have taken a wrong direction as a right.

In the afternoon of the 17th the horizon cleared up a little; and, although the wind was from NE. but moderate, the current being in our favour, we got under-sail, having at last got a view of Cape Possession, within which bay we had lain; and all the rest of the morning we continued turning to windward, with the tide still favouring us. It was not without trouble and anxiety, that, after all, we got clear of the strait being obliged to pass between the Cape de las Virgenes, and a shoal or sandbank, formerly discovered and pointed out by Sarmiento. The wind had drawn round to the N. and NNW. and then all at once passed to the SW. with great fury, attended by lightning and rain. We were thus again reduced to the necessity of sounding, and to drive under bare poles, for the violence of the wind. At $9\frac{1}{2}$ P.M. we had a glimpse through the dark clouds of Cape de las Virgenes, which, at $1\frac{1}{2}$ in the morning following, lay to the north of us. The wind now abated so much that we were able to set some sail; so that, about break of day, on Saturday the 18th of March, we enjoyed the unhoped-for happiness of being entirely out of the Strait of Magellan, and on our way for Spain.

SECTION IV.

Return to Cadiz.

It has been said, that, at $\frac{1}{4}$ past 1 in the morning of the 18th, we were in the mouth of the Strait of Magellan, N. and S. with Cape de las Virgenes. The high wind from SSW. which had forced us to run out with such disorder, cleared the atmosphere, and the moon being remarkably bright, we easily discerned the land, and were able to form our course, which, some hours before, was a very difficult point, both on account of the violent winds and the darkness of the night, and of the shoals which Sarmiento lays down to the eastward of the mouth of the strait. However, having escaped so many dangers as surrounded us some time before, our situation in the mouth, which was about 7 or 8 leagues over, did not appear to be so very hazardous.

At $2\frac{1}{2}$ A.M. being due E. from Cape de las Virgenes, 7 or 8 miles off, we found 25 fathoms fine sand, from which point we steered away NE. 5° E. the wind having veered round from W. to NW. by W.

The morning of the 18th was clear and delightful. It is impossible to describe the joy which prevailed throughout the ship's company, when they saw themselves at last delivered from the imminent dangers with which, for the space of three months, they had been surrounded, and from which they had scarcely hoped ever to escape.

The seaman, who seldom cultivates his mind or elevates his heart, by extending his views beyond the enjoyments and sufferings of the present time, receives no consolation, in his labours, from the consideration that, from them, may result assistance and advantage to his fellow-creatures who may come into the situations through which he passed. But this reflection afforded great delight to our officers, who observed also, that the general good health enjoyed by their men promised a successful voyage to Europe. In the whole crew there was only one man whose case required much attention, on account of a venereal complaint he had contracted before we left Spain; for, although there were five others indisposed, four of them with fevers, in consequence of cold, and one with an attack on the lungs, yet their complaints were of little importance.

On the 19th and 20th, we saw vast numbers of whales, of which, with proper implements and arrangements, many might have been taken.

Till the 21st, we had a very heavy sea from the NE. to N. drawing more and more to the latter point as we got farther out from the land; from which we concluded, that the NE. winds we had met with in the Strait of Magellan had been part of the strong north gales so frequently found in this part of the Ocean. This day, at 9 P.M. we saw a meteor, which much surprised us; for the weather was not of that kind that we should expect lightning. It ran along parallel to the horizon, throwing a greater light than the brightest flash, and lasted about four or five seconds.

On the 23d, after some hours of calm weather, a NE. wind set in; and then veering round by the E. it settled from SE. to SSE., and blew very hard for three days together. The heavens were so overcast, that we could make no sort of astronomical observations; so that, if we had not considered the vessel to be 80 or 90 leagues from land, we would have been very uneasy. But, as we then were, such winds were of great service to us, and greatly forwarded our voyage.

We now altered our course a little to NE. by E. in order to pass about 100 leagues to the E. of the mouth of the River de la Plata, and in this direction ran through the parallels of latitude from 45° to 41°; in crossing which, on our passage to the strait, there had been so great an error in our reckoning, which we accounted for by the setting of the water towards the NE. quarter. At present, however, we had no proper opportunity of examining whether the same circumstances took place; for the wind was very inconstant, and the only altitudes we could observe by the time-pieces were in S. latitude $47\frac{1}{2}°$ and 42°, in which interval of three days the wind continued from the SE. The results of these observations, and of the altitudes at noon, produced a difference of twenty-five miles to the NW. by N. with the reckoning. But this difference, which is very small when compared with that discovered in our outward voyage, between the parallels of 45° and 41°, cannot be depended on, as the reckoning was very uncertain; and because the error between latitude 45° and 42° was in some measure compensated to the westward by the error observed between $47\frac{1}{2}°$ and 45°. From observations made from latitude 42° to 41° on the 25th and 26th of March, we discovered a difference of thirty-six miles in the direction north-easterly, as in our former passage through these parallels.

The wind died away entirely on the 26th, so that the swell which remained became extremely troublesome; for, had any accidents happened to our masts, we must have run for Rio de la Plata, having no means of repairing the damage on-board.

On the 27th, in the morning, we had the wind fresh from NW. and, on the two following days, we crossed the mouth of that river, at the distance of about 100 leagues from the land, steering NE. We perceived currents setting-in towards the river, as we had before observed in passing from N. to S.

As we have not here set down the mean direction of the winds and of our courses, which were all inserted in their proper Tables, we will just notice, that the winds from the eastward obliged us to keep close to them, standing to the northward for three or four days; and, when they came from the NW. we stood towards the NE. without being able to keep the direct course we had proposed, and employed four days in running down from 25° to 24° S. lat.; after which the wind set in from the SE., and, on the 11th and 13th of April, we had a sight of the island called Trinidad or Ascension.

In the French charts are laid down, in the parallel 20° 30′ S., two islands, the most easterly of which is called Trinidad or La Trinité, and the other Ascension; with a remark at the same time, that some navigators assert that the two are only one and the same island, laid down in that distant quarter of the globe according to several different accounts.

Admiral Don John de Lángara, who, by order of his Catholic Majesty, went to examine this point, determined that there was but one island, to which he gave the name of Trinidad, because there is another island called Ascension, in S. latitude 8°, and placed it, agreeably to several series of lunar observations made by himself and other officers of skill and accuracy, in S. latitude 20° 29$\frac{1}{2}$′ and in longitude 24° 13′ W. from Cadiz. According to our position, laid down in the chart, as we lay with respect to this island, we were, at noon of the 13th of April, in W. longitude 24° 20′ 30″, by our reckoning, in 23° 12′, and, by the time-piece, in 22° 58$\frac{1}{2}$′, which last was 1° 22′ to the E. of our place on the chart; on which account we suspected that some alteration had taken place in the time-piece, and, on examining the results indicated by it at noon for some days past, we discovered that, in part, it had undergone some change in its motion.

In twenty-four hours, No. 16 of Berthoud had gained on No. 71 of Arnold 1′ 15$\frac{1}{2}$″; which, even supposing that No. 16 had returned, now that we were in a warm climate, to its former usual advance of 52$\frac{1}{2}$″ on mean time, would indicate that No. 71 had gained in one day only 20′, when we allowed it to gain daily 26″, or mean time. It thence followed, that No. 71, going slower than we supposed, pointed out less time elapsed at Cadiz; consequently, as we were to the west of that meridian, it gave less longitude than it ought, as we discovered, on comparison with our plan on the chart. This was the greatest difference we observed on comparing these time-pieces, Nos. 16 and 71, in any one day; those of the seven preceding days being much less. Nos. 15 and 16 of Berthoud had now returned to their respective differences in those latitudes, which they preserved without variation for many days together,—another proof that the remarkable alteration had taken place in No. 71 alone.

We had noticed for six days before very considerable differences to the eastward, between our reckoning and our position, as indicated by No. 71, which had not occurred on our passage outward through the same latitudes. The heavens had been very cloudy, excepting the four or five days immediately following that of our departure for the Strait of Magellan; so that we had not been able to make observations of lunar distances, in order to examine the state of that time-piece,—an indispensable precaution for receiving satisfactory assistance from such machines, which often alter their movements at the moment when it is least expected.

On the 14th, we were obliged to sail close by the wind, which came away from N.E. to E.; we were, nevertheless, able to make our course good to the NNW. 5° W. for five days together, until we got, on the 17th, into S. lat. 16°, in a position which rendered it absolutely necessary either to sail N. by W., or to make a tack to the eastward, in order to weather Cape St. Augustine, or the coast of Brazil,—a thing which must have happened to few vessels that passed within sight of Trinidad. We were persuaded that, in these latitudes, we should certainly meet with the winds from the SE. quarter, as usually is the case, with clear pleasant weather; but the best judgments in such cases, in these parts, even when best founded, are very uncertain. The heavens continued to be very cloudy, with squalls and rain, as if we had been in the proximity of the Line. On the 18th, however, the wind came round, as we desired, to the SE. quarter, and the atmosphere began to clear up; and, although on the 19th it returned again to the NE., with squalls and showers, yet it soon came back to the former point. On the 21st, breezes set-in from the ESE. with a clear sky: the following day we saw a small vessel, to which we gave chace; but, observing that we did not gain much upon her, and that we were driven out of our course, we gave over the pursuit, and returned to our former point. The sailors, who for a long time back had been without tobacco, had flattered themselves that they would be supplied from the first ship we should meet, especially if she came from Brazil.

One of our officers, who, believing our expedition might last a couple of years at least, had made large provision of Seville snuff, had frequently relieved their wants with it, which they used to wrap up in paper, having first moistened it, and to smoke as if it had been tobacco in the leaf; but, as there remained of his stock only what was requisite for himself for five or six weeks, that resource was cut off. No sooner had we abandoned the chace of the vessel, than we perceived how much these poor fellows were mortified; which, joined to their daily complaints and impatience for some time past, made us feel that it would be highly proper, especially in long voyages, to take on-board a stock of tobacco on public account, to be distributed among the seamen and marines, according to some fit regulations. By this means, much contraband trade would be prevented, as well as the enormous extortion practised in the sale of that article on-board, it being impossible for the commander or officers to put a stop to such abuses, as those very persons who are the objects of such extortion, are always the last to acknowlege that such things are done. Any one at all acquainted with the character of a seaman, knows how much he will sacrifice to the gratification of even the feeblest of his desires and inclinations.

On the 23d we had frequently heavy rains, with little wind; but a breeze coming away more to the southward, we were able to steer NE. by N. in order to pass to the eastward of the island of Fernando Noroño (Noronyo).

On the 24th, we succeeded in making good observations of three series of distances of sun and moon. From the island Trinidad we had counted our longitude by No. 16, taking it for granted, that that island was correctly laid down by Admiral de Lángara, and estimating its daily advance on mean time to be the same as when we left Cadiz, since its difference with No. 15 corresponded with what was observed in the same latitude on our voyage eastwards; but, taking the medium of the lunar distances from the sun, whose extreme did not differ half a degree from each other, it resulted that we were now 47' to the eastward of No. 16, and, consequently, either that the island of Trinidad is more to the east than was formerly stated, or that the time-piece advanced much more than we imagined. In this dilemma, we chose rather to attribute the error to No. 16, and at the same time remarking, that No. 71 of Arnold had now returned to the respective differences with the other two time-pieces which it had maintained prior to the alteration in its motion, we again resorted to it, giving it the preference over the two others of Berthoud, and allowing it the same daily advance on mean time which had been formerly ascertained. The experiment we made on this time-piece was, to calculate the longitude by it, as if no alteration had taken place, and to compare it with that deduced from the medium of lunar distances, whence it appeared, that the latter was 1° 4' to the east of the former, evidently showing that some change had happened in the time-piece. It must however be mentioned, that this alteration was discovered only in consequence of comparisons made after we came within sight of Trinidad.

These combinations were, no doubt, inexact, and might without hesitation have been abandoned; but our anxiety to arrive at the truth, led us perhaps to assign to them an importance to which they were not entitled. When time-pieces once come to undergo alteration in their movement, it is impossible ever to discover with precision their rate of motion, while we are continually changing place, and are destitute of a fixed point of comparison. Besides, we were induced to keep an account by the time-piece, because there had been a great difference westwardly, for the last three days, between our reckoning and the truth, as is experienced by perhaps every vessel within the Torrid Zone, and in the neighbourhood of the Equinoctial.

At day-break of the 25th of April, being in S. latitude 4° 45′, and W. longitude 23° 40′, happened the full-moon, when we stood away NNE. 5° E. that before night we might reach the parallel of the Island Fernando Noroño. When it grew dark we had seen no land; and, although we had not attained that parallel, even with our glasses, we resolved to continue our course, being satisfied in the accuracy of our calculations, as well as of the position assigned to that island of Don John de Lángara and Captain Cook. The latter places it in 26° 18′ W. longitude from Cadiz, and in S. latitude 3° 53′, referring his observations to the highest hill of the island, which appears like the steeple or tower of a church; and we were then full three degrees to the eastward of its meridian.

Night came on very cloudy; however, in some intervals of clear weather, we observed the meridian altitudes of some stars, to discover if there was any remarkable error in our reckoning; and, at $11\frac{1}{2}$ P.M., concluding that we had passed the parallel of the Island of Noroño, we continued our course in the former direction. The clouds dispersed without either squalls or rain, so that this night was the finest we had seen since we were in lat. 30°.

On the 27th, in the morning, we had squally rainy weather, and we steered NNE. to make the most direct course. Next day we had calms and gentle breezes, when we saw several bonitoes, and caught three small ones,—a confirmation of the common opinion that these fish are only to be seen in such weather.

Notwithstanding that the weather, since we left the Strait of Magellan, had been so rainy and inimical to the health of our people, who were almost constantly quite wet, yet, in our sickward there were only three patients; of whom, the worst was one with a venereal infection, who had been at the point of death in the strait, and whose complaint, now complicated with scurvy, rendered his recovery doubtful, at least, during the voyage; nevertheless, he stood it out. The other two had symptoms of the latter terrible malady.

During the 29th we had more calms and heavy falls of rain, now beginning to experience the ordinary weather of the Line, which did not last long; but at noon sprung up a wind from

SE. and, at 2 p.m., we crossed the Equinoctial, about $22\frac{1}{2}°$ W. from Cadiz, according to the time-piece No. 71, for our reckoning was $1\frac{1}{4}°$ more to the eastward. This point was at least $4°$ to the westward of what is laid down in most ships' reckonings; but the observations we made, on crossing the Line to the southward, ought to have greater weight in this case, when the same interruptions are not to be apprehended as in the other. Navigators should therefore cross where they can, and not expose themselves to be detained in these seas, which may be attended with very dangerous consequences.

This night we were obliged to sail with great caution, on account of the rock of San Pedro, whose situation is so uncertain. We wished much to have passed it in the day-time, that we might, by means of lunar distances, have ascertained its position. Many persons doubt even its existence; but several of our crew, amongst whom was one of our under-pilots, asserted that they had seen it on former voyages.

On the 1st of May, at day-break, we descried a vessel to windward coming down on us: at 8 a.m. she hoisted English colours, to which we answered with the Spanish, and lay-to to wait for her. At 10 a.m. an officer went on-board her, who reported that she came from the coast of Angola in Africa, with four hundred negroes, and was bound for Barbadoes. She was called the Fanny of London, Thomas Smith, master, who considered himself to be in $27°$ longitude W. from London, and therefore two or three degrees to the eastward of us; of which we gave him notice. He could furnish none of the articles we wanted, only some tobacco, of which, with the greatest cheerfulness, he supplied us with fifteen or twenty pounds, which was immediately distributed among the crew. We enquired if he stood in need of any assistance from us, to which he answered in the negative; and proceeded on his course. When night came on, he was no longer in sight.

We continued to stand to the northward, with the wind from E. to NE. and NNE. At night the sky was very cloudy, with constant rains and squalls of wind from time to time. The following day passed in calms and rain, until 9 p.m. after a very heavy shower, when the wind set in from E.; we therefore ran N. suiting our sail to this rainy squally weather.

On the 3d, towards evening, we saw a number of white birds flying against the wind, but they were at too great a distance for us to discover of what kind they were; only we can say, that they are not commonly met with in these parts of the Ocean; and, two days afterwards, another small bird alighted on the ship, when we were in N. lat. $10°$, and W. long. $20°$. It had, no doubt, been forced off the land by the late violent winds; and, to avoid its fate, had only the chance of going along with us, or of making the islands or coast of Africa.

The medium of four series of distances of the sun and moon, observed on the 6th and 8th days of May, whose extremes were within eight minutes, being all that the weather had permitted us to make in this quarter of the moon, placed the frigate 19′ to the E. of the position indicated by the time-piece; and three series of distances of the moon from Regulus and Antares, observed in the night of the 10th, gave twenty-six minutes in the same direction.

This evening we saw a bird resembling a pigeon, of an ash-colour, having two forked feathers advancing much beyond the rest of the tail.

In proportion as we advanced in the northern hemisphere, we found the winds to draw round very slowly from NE. to E.; and, on the 12th, in N. lat. 20°, and W. long. 33°, we had them for the first time from the SE. quarter, which enabled us to amend our direction, steering NNE. 5°. N.

On the 14th of May, in lat. 23°, and long. 32°, the winds came from S. to SW.; which gave us the more pleasure, as such winds, in this time of the year, could hardly be expected. We then stood NE. by N. in order to pass to the W. and N. of the Azores; by that means making sure of our way, and not caring to sail to the S. and E. of those islands, although it was a more direct course, lest we should fall into the calms which in summer are generally found in that quarter.

As we were ignorant of the political state of Europe since our departure from Spain, we made preparations for our defence, in the event of any hostile rencounter, exercising the men daily to the use of the great guns. We were, in particular, anxious to learn whether or not the truce lately concluded with Algiers had been broken; but the English vessel we spoke with in the Line, could give no information on that head.

On the 15th, and following days, we saw floating past us several quantities of sea-weed, of a kind which, according to M. Frezier, abounds on the coast of North America, different from that which grows in the Canaries and east coasts of the Atlantic; from which appearance we inferred that the west winds had prevailed in this quarter.

On the evening of the 16th we gave chase to a vessel a-head; and at 11½ P.M. got near enough to speak with her. It cost some trouble to carry on the conversation; and all that we could learn, the captain speaking nothing but English, was, that he came from the coast of Africa, out thirty days, and bound for Liverpool. We soon outsailed her, and returned to our former course, from which we a little deviated, in order to meet with her.

The wind now coming away from the SW. quarter, forwarded us so much, that, on the 17th, at noon, we were come to N. lat. 27° ¼, and W. long. 30°, when they began to blow from the northward; and, in the night of the 19th, after we had been becalmed almost the whole of the day, they sprang up from the E. and so continued till the 23d. We then stood to the northward, and came into lat. 34½°, and long. 29° 40'. After some hours of calm, we had the wind from the SE. quarter; which, on the 24th, changed to the SW. when we again stood for the N. This day we compared thirteen series of distances of the sun and moon observed in this quarter of the moon, the medium of which gave our longitude 36' more the eastward of the timepiece, with which difference we corrected the account kept by it. It appeared from these results, that No. 71 of Arnold had very accurately pointed out the course we made good, and the daily errors of our reckoning.

In taking these observations of the distances between the sun and the moon, we made use of the circular instrument, or circle of repetition, of the Chevalier Borda, which proved to be excellent in its kind, and whose properties are so well pointed out by M. Jacinto Magellan (Magallanes), in one of the *Memoires* he has published on the use of astronomical instruments, (Paris, 4to. 1775;) who, in the same work, mentions his reasons for believing that he had employed similar instruments a considerable time before he had heard any thing of the experiments of that astronomer.

At 9 P.M. being then in N. lat. 37° 21', we ran to the E. in order to keep clear of the Vigia, which, although it is laid down by M. Verdun de la Crenne, in his chart of these seas, in latitude 38° 13', he says also, that perhaps it ought to be placed in 37° 30'; in which latitude he has laid it down a second time. Had we trusted to our reckoning, we would not have taken this course, which would afterwards have prevented us from passing to the westward of the Azores with the winds which generally blow from W. to NW.; but, according to our observations for determining the longitude, in which we placed greater confidence than in our reckoning, we were still four degrees to the W. of those islands.

The pilot of the ship Buen Consejo, on his return from Lima, had some conversation at Fayal, in the island of Flores, one of the Azores, with a Portuguese pilot of that island, who gave him the following information: viz. "The Vigia lies WSW. from the island Flores 66⅔ leagues. An English vessel from

Madeira was lost upon it; and the crew in the long-boat steered ENE. the above distance, which brought them to Fayal. This pilot himself proceeded, in consequence of this information, to examine the Vigia, and accordingly met with it, running back the way the boat had come. It has two reefs of rocks, one about 180 yards long on the WNW. part, and another of 100 yards to the south, with a cluster of rocks, covered at high water, and a small sandy beach of forty yards long, to the WNW. of these rocks." This information, which, notwithstanding it is so circumstantial, seems to want authority, is copied from a note of one of the journals of our second captain.

As vessels coming from the south commonly pass in the neighbourhood of this spot (la Vigia), whose situation is so undetermined, they are obliged to be much on their guard, from N. lat. $37\frac{1}{2}°$ to $38\frac{1}{3}$, lying-by for one or more nights together;—a great hinderance to navigation, which demands that a point so interesting to all maritime nations should no longer remain in uncertainty.

On the 25th we had easy variable winds from the NE. quarter, changing on the 26th to SW., which carried us into lat. 40°, and W. long. 28°; from whence we steered eastwardly; and on the 29th, in the morning, came in sight of the island Cuervo, one of the Azores. At noon, according to the latitude of this island, laid down in the chart of Verdun, we were in W. long. 24° 29'; according to the time-piece, No. 71, by which we kept our account, in 24° 26'; but, by the reckoning, in 19° 18', and, consequently, 5° 11' to the E. of our place of markation in the chart,—a space which, in this parallel of latitude, is equal to seventy-nine leagues.

The time-piece differed, in three days, 52' to the W. of the reckoning. Three bases measured by the log in the morning, in order to calculate our distance from the island, produced results so distant from what they certainly ought to have been, that we were convinced of the existence of a strong current, whose direction and force explained the differences in our accounts.

The night was calm, and next morning the wind set in from S. to SE. with cloudy weather; which, continuing for five days together, forced us, much against our inclination, to run to the northward as far as latitude $43\frac{1}{4}°$;—a circumstance the more painful to us, that, arriving now in a cold moist climate, the scurvy began to make its appearance among the ship's company, particularly in the commander, who, notwithstanding his great spirit, was so ill as to be confined to his bed.

We were also by these winds driven into the parallel of the

other Vigia, in latitude 42° 30';—a danger we had before had no thought of incurring.

As the heavens were in general overcast, with but few intervals of clear weather, our time-piece was of great use in observing meridian altitudes of the sun, when we had calculated the hour of his southing; and by this means we had great confidence in our calculations of the latitude, which, without such helps, must have been very uncertain. The assistance to be derived from those machines ought highly to increase their value in the opinion of all navigators.

On the 2d of June the wind began to draw towards the W. which enabled us to improve our direction, and thereby to diminish our latitude. The following day we discovered a vessel, and at $2\frac{1}{2}$ P.M. were within speech of her. She proved to be the Na. Sra. de la Antigua, from Brazil to Oporto, had been ninety-seven days on her passage, and had not seen the Azores. We set the captain right as to his position, in which he was very much mistaken.

The night of the 4th was spent in calms, but in the morning the wind sprung up from the eastward. We now began to discover a number of vessels standing to the northward. The meeting with them in this latitude, at such a distance from the land, proved to us that the winds had blown from the N.E. quarter for some time, as we afterwards were told had been the case; on which account we were glad that we had passed to the W. and N. of the Azores.

In these two last days we took eight series of lunar distances from the sun, the mean of which placed the frigate 6' 27" to the eastward of the position given by the time-piece; and, applying this small correction to our account, we steered so as to fall in fifteen or twenty leagues to the west of Cape St. Vincent, paying no attention to our reckoning, which, in these few days past, had contracted an error of $1\frac{1}{2}°$ to the E.; which, had we trusted to it, might have produced very disagreeable consequences.

We continued our voyage until the 9th, without any remarkable occurrence, only seeing sundry vessels from time to time; and, at 5 P.M., at last came within sight of the land. In half an hour more we could discern it to be Cape St. Vincent; and at 10 P.M. by moon-light, we observed it to the N.

Our reckoning on to this period placed us in longitude 54' W. from Cadiz, whereas that of the Cape is 2° 43'; so that the error in the reckoning, after a run of eleven days from the island Cuervo, had been 1° 49', or twenty-nine leagues, more to the eastward than the true position of the vessel. According to the time-piece, the error of the reckoning was 1° 57½' to the E.; therefore, the true position was 8½', or two leagues and a half, to the E. of the time-piece.

Although, when night came on the 10th, we had not got sight of Cadiz, nor could learn the exact point on which it lay from the crew of a fishing-boat, which supplied us with a little fresh bread and fish, we began to sound for the channel leading to the bay, and at 10 P.M. we discovered the light-house of San Sebastian; at 11¼ P.M. we came to anchor NW. by W. from that light-house, in twenty fathoms sand, &c. Next morning (the 11th June) we moved into the Bay of Cadiz, and came to anchor, after a voyage of eight months and two days, without having lost a single individual of our company during the whole expedition. The commander, and two of the crew, were the only persons seriously ill, and other sixteen slightly affected with scurvy; all of whom were in a short time perfectly restored to their ordinary state of health.

We have already taken notice of the death of two seamen, one on our way out, and the other in the strait; but we do not count on them, as their complaints were contracted on land, before we sailed, and were in no manner occasioned by the voyage.

Those only who return to their native country, after an absence, and such an absence as ours was, of eight months, can conceive the pleasure we felt, at once more meeting with our friends and countrymen at home; the commander and officers rejoicing that they had so successfully accomplished their undertaking,* and the seamen, who quickly forget their dangers and toils, with the satisfactory reflection, that they had strength and courage to resist and overcome all difficulties:—the expedition to the Strait of Magellan manifesting to all what the vigour and steadiness of the Spanish seaman are able to perform.

---

* From which some benefit to the world might arise.

PART II.

SECTION I.

Description of the Strait of Magallanes.—Division of the Country into high and low.—Temperature: Qualities of the Soil.—Productions of the Strait: Herbs, Plants, Flowers, Shrubs, and Trees.—Description of the Quadrupeds, Birds, Fishes, and Insects.

The opinion of the greatest part of naturalists is not improbable, that this Strait has been formed by the earthquakes and effects of the volcanos of this part of the globe. M. Buffon, in his " *Epoques de la Nature*," thinks that the high mountainous part of the country is very ancient, and that the plains are comparatively modern; assigning as a reason for this opinion, that the sea, agitated by the winds, constant and violent from the west, gradually consuming the west coast of the continent of South America, has gained on the land, on that side, as far as its power was able to prevail; from which he infers, that the land now seen on that west side must be very ancient; and also that, on the contrary, the sea loses ground, or falls off, on the eastern coast, leaving uncovered and visible such low lands as are now seen near Cape de las Virgines: so that the low land extending north from Point de Micra to the ridge of hills reaching behind Cape de las Virgines to Cape Possession, is all very modern; and that, in former times, the sea extended over that low land up to the above ridge, which was then a high steep shore.

We will, however, now leave these conjectures, and begin to treat of such objects as have been observed; and the first thing to be noticed is, that the country in the vicinity of the Strait of Magellan must be considered under two different points of view, separating the low or plain part from the mountainous; since there is a total difference, not only in their natural qualities and productions, but also in their inhabitants.

The plains, or low country, occupy all that part of the continent on the north side of the strait from Cape de las Virgines westward to Cape Negro, but it is not easy to ascertain its extent towards north and east; only we may be certain it reaches a great way in that direction, and that it joins with the vast plains *(pampas)* of the province of Buenos Ayres and the Patagonian coast, from which it seems to have no sensible difference.

On the south side of the strait, the Tierra del Fuego extends westward, from Cape Espiritu Santo, as far as Cape San Valentin, and to the south-east, according to the account of the Nodales, as far as Cape de Pinas, where the ground begins to rise up, and to become mountainous: so that that portion of Tierra del Fuego lying between the channel of St. Sebastian on the south, and the Strait of Magellan and the channel of S. Maria de la Cobeza on the north and west, may be considered as one great field of low land, different in every respect from those islands properly called the Tierra del Fuego.*

From the above-mentioned Cape Negro on the main land, on to that of Victoria, at the west extremity of the strait, the continent presents only a group of barren mountains, with some plain ground at their bottom, which are the beginning of the famous Cordilleras (chain) of the Andes, which divide South America into eastern and western, running through it, north and south, for the distance of 1,700 leagues.

This cordillera begins at the most southerly point of the north coast of the Strait of Magellan, which is the Moro de San Agueda, otherwise called Cape Forward, which may be considered to be the southern extremity of that vast continent, whose northern limits are still so uncertain.

Along the coast of Fuego also, from Cape San Valentin to Cape Pilares, are seen pinnacles of prodigious height, whose appearance is, if possible, still more horrid than that of the mountains of the continent; and showing, at first view, that that part of the country is nothing but a group of islands,—a manifest proof of the revolutions which our globe has undergone.

The track of country which we distinguish by the appellation plains, or low lands, is not so even as not to have sundry inequalities formed by little hills, which occur so frequently, that no considerable portion of the surface is free from heights and hollows. In both these, the nature of the soil is of the same quality, being a compound of darkish-coloured sandy earth; at least, it is so on the surface, and we had no opportunities of digging into it. Nevertheless, from what we saw, in such places as the ground is cut into along the shore, there seems to be no

---

* This tract of South America was not so called, the *Land of Fire*, from any extraordinary heat experienced in its neighbourhood, but from the *fires* lighted up a' ng the coasts, when the first navigators were seen in those seas.

other difference than only the addition of some small stones. It appears also, that this soil contains a quantity of very acrid salts, which oppose the vegetation of plants and trees; there being on it but very few of the former, and not the smallest vestige of the latter.

As we had no occasion of visiting any part of the Tierra del Fuego, we can only say what appeared to our view at a distance; which is, that it seemed to be in all respects similar to the continent, with this difference only, that it is more broken and uneven; so that, in that point, it has more resemblance to the Falkland Islands, and that, on that account, it is but probable that the productions of both should be much alike.

So different from this plain country is the aspect and appearance of the mountainous track to the westward, that it seems impossible that nature, which, in her changes, generally preserves a certain gradation, should here make so sudden an alteration.

It is but natural to suppose, that the lofty mountains which occupy this tract are all of the same qualities; but it is difficult to ascertain the nature of the soil of which their sides, and the narrow levels at their feet, are composed; for these parts are either entirely clothed with uncommonly thick forests, whose dead trunks and branches, with other decayed vegetables and shrubs, have formed a surface much raised above the true ground, or they are covered with a kind of plant resembling esparto, (a species of rush, greatly resembling *bents*, very common in Spain,) but much more brittle, from a palm to half a yard in height; and its colour, when grown up, is like that of esparto when dried, or dead.

The mountains are covered, in general, with trees, for two-thirds of their height, and the remainder is nothing but a mass of naked, barren, and rugged pinnacles, of a reddish colour; although there are some parts of a different nature,—that is, of common granite, called by naturalists *saxum*, which they consider to be the heart, or primitive rock, of all mountains. The upper parts of these ridges are commonly overspread with snow and ice; which, on account of the extreme humidity of the other parts, is dissolved soon after it falls. We observed nothing particular on the summits of such mountains as we examined, and they appeared in all respects to correspond to the description given by Don A. de Ulloa of the Cordilleras, of which these are a part.

Between Cape Redondo or San Isidro, and Cape Forward, there is a hill very steep, and cut down perpendicularly over the sea, with a depth of water upwards of fifty feet close to the foot of it, clothed with fair green trees all over the summit, which seem to be entirely composed of shells and other petrified substances; on which account M. de Bougainville called it Cape Remarkable.

The sole difference between these mountains and those of the Tierra del Fuego, is, that they are not so well covered with wood; nor are the trees there so large or vigorous, being also in general much more loaded with snow.

Our utter unacquaintance with minerals may perhaps have been the cause of our meeting with no traces of them in these parts; yet the natives (Indians) often brought to us pieces of a stone with which they light their fires, which they said were found in the mountains; which stones must, no doubt, contain some kind of metal, as we imagined, from the specks of a substance more hard and brilliant than the rest of the stone. When it is struck with the steel, it gives fire, and smells like sulphur; from which circumstance it is probable, that minerals of different kinds might be found in the bowels of these mountains,* and that these vestiges indicate the existence of volcanos, in former times, in this part of the globe.

Although little rain fell during the fifteen days of our residence in the plain part of the strait, yet the dryness which we noticed, seemed to be occasioned more by the sandy, and consequently uncompact, nature of the soil, than by the want of rain or dew, which, when they fell, penetrated through it so speedily, that, soon after a shower, it could not be perceived to have rained at all; to which must be added, that the prevailing winds in that quarter are in themselves dry and violent, as might be perceived from the plants, which are all laid over in the direction of these winds: on which account, the soil does not appear proper for the cultivation of any European grain, as has been, after many trials, found to be the case in our settlements in the Falkland Islands, of which the soil is of the same kind.

In all this plain tract of country, we found no river or brook deserving notice, only some trifling channel, almost without water: but, on the other hand, there are several ponds, or small lakes of fresh water, which serve to supply the inhabitants. Of its qualities we can say nothing, as we never used any on-board, on account of the difficulty of procuring it in any useful quantity.

It is not easy to ascertain the temperature of the climate of this part of the strait, from the short stay we made in it; for, as

---

* *Note of the Original.*—Pedro Sarmiento pretends that this stone is the ore of silver or of gold *de Veta*, as it entirely resembles the *curiquixo de porco del Peru:* These are his proper terms.

the sun was then about eighteen hours above the horizon, it would be improper, from observations at that time, to infer what it might be at others. However, even in this season, an excellent English thermometer with mercury, graduated according to the scale of Réaumur, exposed constantly to the air in its case, never rose above nine degrees, and sometimes only to five degrees; from which we may conclude, how cold it must be in other times of the year, especially considering that the winds from the W. and WSW. passing over mountains covered with eternal snows, and thereby loaded with cold frozen particles, must greatly enhance the severity of the cold. The heavens are generally clear, and the atmosphere bright; at least, they were so on the two occasions of our passing that way: but this must be understood particularly of the Cape de las Virgines and its neighbourhood; for, even so soon to the westward as at the first pass of Esperanza, the proximity of the mountains begins to be sensibly felt, the atmosphere there being but rarely free from vapours.

The temperature of the mountainous track is different in different parts. From Cape Negro to Cape Forward it is the most mild, and the appearance of the country the most agreeable. From the latter point to the Cape of St. Geronimo, the climate becomes more severe, and the face of the country more rude and unpleasant. But, even this is nothing to be compared with the remainder of the strait on its west point at Cape Vittoria; to which last district, Narborough, with great reason, gave the horrid appellation of the *Desolation of the South.*

In the midst of the summer of this part of the earth we experienced severe cold, and a singular inconstancy in the weather: very seldom did we enjoy a clear sky, and short were the moments in which we perceived the heat of the sun. Not a day passed without some rain; and, in general, it did not even intermit to rain the whole day long. The thermometer stood at 6° and 7°, and often it fell to zero. At the same time, it must be remarked, that the mountains with which our vessel was surrounded, must necessarily have diminished considerably the cold, whose severity we found to be most intense, and almost insupportable on their summits.

It cannot be doubted that the steep, lofty, and barren, rocks and pinnacles of this part of the strait, covered with perpetual snows, presenting an aspect equally gloomy and frightful, contribute much to the humidity and cold of the atmosphere; for which reason the air is constantly loaded with vapours and fogs so dense, that often the most furious hurricanes are not able to dissipate them. If here, as in other parts of the world, the cold should increase proportionably in the winter season, it must be next to intolerable.

We had no opportunity of experiencing this severity; but the Dutch, who were detained in this strait by contrary weather, being obliged to remain all the winter in the Bay des Cordes, lost no less than 80 persons out of their number by the inclemency of the climate: although we need not have recourse to foreign examples, when we reflect on the fate of the colonies planted by Sarmiento, which were entirely destroyed by the climate.

All authors agree, that the southern hemisphere is, in equal latitudes, twice as cold as the northern. Some pretend, that this arises from the greater space of the former which is occupied by the waters of the Ocean; from whence it comes, that, in certain seasons, banks and shoals of snow and ice are met with in no very high latitudes; and hence also proceed the winds, which continually blow with violence from the westward, which, passing over an immense extent of ocean, without meeting any obstacles to interrupt or divert their course, gradually acquire such force, as to be capable of occasioning the most dreadful effects; in particular, rendering the passage from the north and east, round Cape Horn, so painful and dangerous. In the Strait of Magellan we observed some variations in the winds; only, in general, we found that they followed the direction of the channels amongst the islands and mountains; and the impression produced by an atmosphere so loaded and confined between such lofty mountains, contributed much to form those furious gales, blasts, or hurricanes, which we sometimes experienced, and whose fury renders so tedious the passage through this strait.

It is impossible to conceive the moisture prevailing in all these parts, and the multitudes of rivulets and falls of water which, precipitating themselves from the higher parts of the mountains, form a prospect, at first view most agreeable; but which, on a nearer approach, soon produces very opposite sensations.

These waters are very good when used immediately after they are taken up; but we found that on board ship they did not long remain so, soon acquiring a disagreeable flavour,—a proof of their bad qualities.

Such being the soil, and such the climate, of the low or plain part of the country bordering on the Strait of Magellan, it is not wonderful that it should produce only the few plants which we are now going to describe; observing, at the same time, they are all found in the neighbourhood of the sea: for we had not occasion to penetrate much into the country; so that it would be wrong either to assert or to believe, that there are not in the interior other classes.

The whole of the district we have described as so unfit for vegetation, from the excessive want of moisture in the soil, is, nevertheless, overrun with a species of plant resembling oat-grass, and with another which grows in plenty in the Falkland Islands, and is there called *paxonol*. In the month of December this was in its full vigour, the colour between green and yellow, as then nearly ripe and drying up, when it remains like straw. This is the substance with which the Patagonians make their torches; and, as far as we saw, it furnished good matter for the flame, being else, in all appearance, extremely fit for the use of cattle, as has been experienced in the Falkland Islands.

1st. There is one plant, two feet in height, very thick and bushy, its leaves like those of the cypress, and of the same colour: at the extremity of each is found a small yellow flower; so that each little branch forms a sort of nosegay or bouquet of flowers, which are very small and of a strong aromatic smell, stronger than thyme. The taste is very bitter and resinous. This plant produces no prickles, nor fruit of any sort: its roots are very much scattered, although they are but slender and weak. When you rub the leaves against the hand, they leave a very agreeable and refreshing odour. It resembles a little the *erica*, or heath, of Spain; but may rather be considered as a species peculiar to this strait.

2d. The next plant has but few leaves, which are small, and covered with a down. It is the shape of the palm of the hand; the colour of the upper surface a bright-green, and white and more downy on the under surface: its taste is somewhat subacid. The stalk is about 1½ foot high, on which it sets out some flowers, which are white in the leaves, but yellow in the centre of the calyx, resembling the marygold: these flowers are always found in a cluster of three or four together; the stalk is also downy and slender; the root, which is white, is from five to seven inches long: in some properties it resembles the sorrel.

3d. The third is about one foot high; the leaves smaller than those of sage, being whitish, thick, and hairy; their smell a little aromatic, and taste bitter: it seems to be a species of *canipitis*, or *semper viva*, of the fields.

4th. The fourth plant is a kind of shrub, little more than one foot high, spreading over the ground for more than a yard in circumference; the leaves round, shaped like the fruit of the almond; the colour a dark-green; its branches thick set with prickles in such regular order, that, under each leaf, they form a cross, on which the leaf rests; the taste is acid but disagreeable, and it has a sensible smell: it bears a small round fruit, of the same taste; but when we saw it, it was not ripe.

These four plants are all that deserve any description. The Abbé Perneti, in his voyage to the Falkland Islands, makes mention of some of these, besides several others, peculiar to these islands. It is not improbable, that an experienced botanist and naturalist might have discovered in this part of the strait greater treasures, to increase the catalogue of plants already known; but in that, this tract of country will always remain one of the most barren and unfit for the production of necessaries for men, at least in the article of vegetables.

If this plain country were proper for the growth of trees, it is but reasonable to suppose that we should have seen some on it; since the violent winds, blowing almost incessantly from the westward, must, in all probability, on numberless occasions, have conveyed hither the seeds of those with which the mountainous tracts to the west are almost entirely covered.

This supposition is supported by the many and useless attempts made by the French and English to raise trees on their respective settlements in the Falkland Islands, transporting them, with all possible precaution, from the Strait of Magellan, which is at no great distance: but neither the one nor the other have hitherto been successful.

When we (the Spaniards) got possession of these islands, in 1764, we also used the most strenuous exertions to the same effect, carrying thither not only young plants, but even the soil, from Buenos Ayres; by which precaution we succeeded in making them take root and live, but not to come to any perfection; and even to procure a few cabbages and garden-stuff, although not in perfect ripeness, which never is obtained, it is necessary to sow them under the shelter of some slope or rising ground, and to surround them with hurdles, to defend them from the winds. The same precautions were employed in raising trees; but no advantage was ever reaped from such expensive labour: all which proves the similarity of soil in these islands with the country we are now describing at the mouth of the strait.

We come now to treat of the quadrupeds of this country, on which subject we must remark, that it appeared very strange, that, in all this tract, we should not meet with the least trace, nor acquire the smallest information, of horned cattle, which have so prodigiously increased all over the territory of Buenos Ayres. Perhaps these most useful animals have never arrived at this southern extremity of South America, on account of the large rivers, and other interruptions, which they have not been able to overcome.

The first animal that presented itself to our view, was the guanaco lamas or lama, of which we might have given a circumstantial description, were it not that such accounts are already common, in all treatises on natural history. We shall therefore confine ourselves to some remarks on the peculiarities of those animals which are found in the neighbourhood of the strait.

The celebrated naturalist, M. de Buffon, is of opinion, that the lama only inhabits the coldest regions of the Cordilleras of the Andes; but this supposition cannot be reconciled with what is related of them by D. A. de Ulloa, in the Account of his Journey to Peru, nor with the great numbers of them to be found along the Patagonian coast and the plain part of the Strait of Magellan, where they constitute the chief article of the food, and the wealth of the inhabitants. In all the different occasions of our intercourse with the Patagonians, the lama was the only article they presented to us, their abundance of these animals exciting our wonder; but, although we saw them so frequently on the shore, we had no opportunity of killing any of them.

It is not to be wondered at, that they should inhabit a tract of country so destitute of water as this is, since it is well known, that they consume but a very small quantity of either food or drink, and frequently quench their thirst by keeping their mouths moist with their own saliva, with which they are furnished in much greater abundance than any other kind of animals.

The lama has often been carried to Spain, but has never propagated, and has lived but a very short time; showing that that animal thrives only in the country from which it originally comes.

The plains abound no less with zorillos (called izqurepatly), whose fur is as pleasant to the eye and the touch, as the smell of its urine is pestiferous, which may be perceived at a very great distance. Our officers shot some of them, but were obliged to throw them into the sea, to prevent the vessel from being poisoned with their abominable smell. Great care is requisite in depriving their fur of this most offensive odour, which would render them totally unserviceable; and, even after they are properly prepared, they must be kept perfectly dry; for, on the least communication with water, they again revive their loathsome smell.

All naturalists agree, that this animal (the zorillo) is found only in the New World; and M. de Buffon has very properly remarked a difference which subsists between those of the south parts of America, and those in the districts of Carthagena and the river *Orinoco* [not Oronoco], both in the size and the colour of the skin, as also in the smell of its urine, which is still more abominable than even that of the zorillo of Magellan.

We have but little to observe concerning the horse, of which the Patagonians make such constant use, as it is well known that the original Americans were totally unacquainted with that noble animal, and his usefulness in the purposes of civilized life, until they had experienced the advantages derived from them by the Europeans, who first transported them to this part of the globe; by that means greatly facilitating their conquest of America.

The perfect similarity proves manifestly that the Patagonians have drawn, and still continue to receive, such horses as they now possess in so great numbers, from the immense plains or *pampas* of Buenos Ayres, where these animals have increased to a degree almost beyond conception or belief.

The author (Walter) of the Voyage of Commodore Anson round the World, asserts, that the natives of these parts prefer the flesh of horses for food to that of any other animal; but, although we took a great deal of pains to discover the truth of this assertion, we never were able to learn whether the Patagonians do really follow that practice: on the contrary, we are inclined to think that it is at best extremely doubtful.

Such faithful companions of these natives are their dogs, that they are very seldom seen without a vast crowd of them. Their race resembles that which at Buenos Ayres is called *cimarrones*, (wild, untamed, also spotted or speckled,) and from whence, most probably, they originally received them, whither they were first carried by the Europeans; since it is certain, what is related by the histories of America, and which is confirmed by Cook, in his first voyage to the South Seas, that the indigenous animals of the canine race, in that country, never bark; whereas, those which accompanied the Patagonians gave unequivocal proofs, and that even at a great distance, of their being descended from ancestors natives of the Old Continent.

As the low parts of this strait are entirely destitute of trees, it is not surprising that few birds should be met with; we shall therefore omit the aquatic fowl, which are common to both parts of the strait, and observe, that we only saw some of those large birds of this continent, which, from their general resemblance to the ostrich, have been distinguished by the same name; but which, when carefully examined, are found to be essentially different from the ostrich: on which account, several Spanish naturalists, as well as Buffon, have reserved to the bird the appellation *tuyu*, by which it is known to the inhabitants of the district in which it was first observed: so that the *tuyu*, which is only found in America, ought not to be confounded with the ostrich, which is a bird peculiar to Africa.

We also saw some birds of prey; amongst others, a kind of small eagle, called by the naturalists the little eagle or grey falcon, which is so frequent in the Falkland Islands.

These are all the animals which are to be found at the entrance of the Strait of Magellan.

We met with no fish of any kind here, so that it must be far from plentiful; and the shores also seem to be destitute of shellfish.

We come now to the mountainous part of the Strait of Magellan, of whose temperature we have already made mention. In the narrow plains and levels along the foot of the mountains grow small round heaps, formed by a plant whose leaves are round, and so closely set and interwoven together, that each plant forms a sort of carpet, extremely equal and even, having on the inside nothing but the roots, which, in proportion as they grow, continue to increase this heap of leaves, until it assumes the appearance of a large round loaf. This plant is called by botanists *sedum minimum*.

These heaps or loaves, as they may be called, are from one to two feet high, and the same in diameter, and, when they are in their vigour, are so strong as to bear the weight of a man; but we observed, that, when they began to decay, they easily gave way, on placing the foot on them. When they are in a middle state, that is, neither so green as to possess all their power of resistance, nor so decayed as to have the roots putrid, they raise and lower their surface, when one stands on them, with a sort of elasticity or tremulous motion, produced by their own strength, as also by the moss or green crust of the ground, which grows up among them.

The surface or soil in which these loaf-shaped heaps grow, is not the solid ground, but only the remains of other heaps of the same kind, corrupted by the moisture; so that the real soil or earth is not met with for four, five, or six, feet lower down, which must render extremely difficult, if not entirely abortive, any attempt to bring into a state of cultivation this ground, which, in all probability, has lain in its present state from its first formation. The proper soil consists of a kind of clay, of a darkish colour, and light, with some small stones, and a little fine sand; so that, if the soil were not so beset with the above-mentioned plant and crust, it seems to be not unapt for cultivation and agriculture, at least, as far as we may judge by the abundance and vigour of all the vegetables which it produces.

The plant just described covers almost the whole surface; of which we have thought it better to give the foregoing description, than to lose time in enquiring for a proper name for it, or a correct assimilation to some other better known, which perhaps would be so much labour thrown away; being convinced that, from what has been related concerning it, experienced botanists will be at no loss to distinguish it, and to arrange it according to its proper class of vegetable productions.

There is another plant in great abundance, near two yards high, very thick set with leaves from the root upwards, which are of a bright-green, and arranged in the shape of a cross; the flowers are white and beautiful, the petals being very small, grouped together like a nosegay. The natives eat this plant, which is to them a great dainty. We had not the good fortune to procure its seed in proper season, which consists of a few square long grains in the flower. The taste of this plant is subacid, with very little sweetness in it.

Another plant is frequently found, whose leaves are of the shape and colour of the vine, but of the size of the ivy. Its height is not quite three yards. In summer it sends out the fruit, which consists of clusters of berries, about the size of a large pea, very black, and sweet; of which our people ate freely, without experiencing any inconvenience therefrom. This plant is the *uva ursæ*, having the same figure and property with others of the same kind that are well known.

Another species of the same shrub, of a yellow colour, is also found here, with a smaller leaf than the former; having on its branches a fruit of the same taste and colour, but a different shape: so that it may be considered as belonging to the same kind of plant, and possessed of the same properties and virtues.

Intermingled with these shrubs is found, on the ground, plants bearing a flower, which, even in Europe, the fair sex might esteem handsome. This flower is bell-shaped, of a rose-colour, growing on a kind of small myrtle. Also another plant, with leaves like the myrtle, produces numbers of white flowers, of delicious smell: this gives a reddish round fruit, like a pea, having within it a stone like the plum. Besides which, there are three other kinds of the same plant. The taste, far from being agreeable, is dry and insipid; but the leaves are harsh and astringent, so that they may be supposed to contain more virtue than the fruit.

Near the rivulets or streams of water, there is commonly found in great abundance, on the ground, a plant, different from the melon, but very much resembling it in the leaf; each of which springs out of the ground by a separate pedicle, the colour of an ordinary green; the pedicle, or foot-stalk, reddish; the taste of the leaf very bitter. We also found this plant in the mountain de la Cruz, having in its calyx a small scarlet berry, like an unripe mulberry; the root is long, but not thick. From the properties of this plant, we called it *Malva Magellanica*.

In one of the bogs we examined we found a great quantity of fern, resembling that which grows in Spain; and, in various other places, a species of maidenhair, but very different from that which grows in moist situations.

There are also along the beach many plants, whose height does not exceed two feet, with leaves as large as those of the white beet.

On the trunks of trees, and beside the channels of water, is found a kind of vetch, or knot-grass, which is a plant with leaves as small as those of lentils; its stalks very broad, and of a dry insipid taste.

Also, along the sea-shore, we met with some shrubs, whose leaves are very fine and delicate, resembling the willow, of a bright-green colour; the flowers scarlet and bell-shaped, having in the centre three small blue petals, enclosing the calyx, altogether offering a very pleasant sight to the eye. The seed lies in a small sheath, like a kernel, but more slender and round; the stem is very crooked, and in general covered with a coat of moss; but the wood is neither strong nor heavy.

Near the beach is also found a large quantity of wild parsley, or parsley of Macedonia, of a tolerably agreeable flavour. On account of the antiscorbutic qualities of this plant, we made constant use of it in our vessel; the ship's company eating it, with much advantage, both in soup and broth, and by way of salad.

In the interior of the woods we met with some plants of anise, but were not so fortunate as to find its grains or seed, notwithstanding that we were in the strait during the season in which it is ripe.

Great part of these woods is overrun with a plant very like rosemary, but which is of a different kind: it is of various heights, the tallest not being two yards; each plant being thick and bushy from the ground upwards; the leaves of a bright-green, whitish in the upper surfaces, and a little downy on the interior; their length about one inch, and greatest breadth one quarter of an inch. It bears flowers on each branch of the upper part of the shrub, of a white colour, but of very little smell; the taste bitter, and somewhat insipid; but, when it is burnt, it exhales a very pleasant odour.

The myrtle is the shrub which produces the fruit which has been used by perhaps all the travellers who have passed this strait, on account of its subacid sweetish taste, being of a cooling quality, and extremely pleasant, when perfectly ripe. Of this fruit there are many kinds; some being round, others oblong, others heart-shaped. They vary also in their colour; some being black, others red, rose-colour, or entirely white; which last are the sweetest.

This plant, called by Sarmiento *montina*, is of different sizes, some being found from one foot to two yards, all producing fruit in their season. The leaves are smallish, but long and sharp-pointed; so much so, that, in getting the fruit, they prick the hands: the colour of all these leaves is a dark-green, and insipid and astringent. This fruit constitutes a part of the food of the Indians, and our crew also eat them abundantly.

Although we saw no more, it would be rash to assert, that the Strait of Magellan produces no other plants than those we have recapitulated, since it is but natural to imagine that it may furnish many other kinds, especially in the mountainous track, of which we examined but a very small portion. It was a fortunate season for our enquiries, when we were in these parts, as almost all the plants and shrubs were then in flower, and in their greatest vigour, which much facilitated our procuring the foregoing account of each sort. We now proceed to that of the trees.

There are three kinds of trees found in the vast forests which clothe almost the whole surface of the mountainous track of the strait. The first, which is the least valuable, has some resemblance to the beech of the northern parts of the world, but still differs very materially from that tree; for which reason, it appears, that the celebrated Mr. Banks, (now Sir Joseph,) who accompanied Captain Cook in his first voyage to the South Seas, gives it the name *fagus antarcticus;* at the same time observing, that he considered it to be peculiar to this country.

The height of this tree is prodigious; but, in many that we cut down, we found the heart damaged, so that it is not fit for work: nor do its fibres possess that strength which might, from its great size, be expected.

The second kind, which is more abundant, is called by naturalists *betula antarctica*, although it bears no resemblance to the birch. These trees are found of all sizes; some so large and straight, that they might form masts, or any other part of the rigging, of a ship, if the timber were not intrinsically so heavy. The timber is of a white colour, and it cleaves or splits only from top to bottom: when the moisture it contains has been dried off, it must be excellent for any kind of carpenter's or joiner's work. The leaves form groups or bunches of different sizes in the branches: their colour a bright-green; their size about that of the nail of the thumb, of an oval shape, terminating in a point, not quite sharp, and indented all round the edge.

The fruit is about the size of a large pea, covered with a very aromatic resin or gum, affording a refreshing smell, when rubbed with the fingers. This resin likewise perspires through the substance of the tree, between the coats of the fruit, and then, hardening with the external cold, remains like a small button or knob on each fruit.

The seed is small, black, and round, like a large pea, having in the centre a black powder, necessary for its propagation and fecundity, which is in great abundance.

The bark of these trees is proportioned to their size, having seen that of some which were not less than thirty-five feet in circumference. It is easily separated from the wood, and the Indians make use of it to form their canoes.

We carried on-board several of these resinous fruits, to see whether any use could be made of their virtues; and, by their smell, they resembled gum copal.

The third kind of tree, which is the most valuable, and not the least common, in this part of the strait, is that which, having been discovered by Captain Winter, on the south and west parts of Tierra del Fuego, has obtained his name: many botanists call it *laurus nobilis,* and the bark Winter's-bark. There are of these trees all sizes. The leaves resemble entirely those of the laurel or bay, about five inches long, and one inch and a half in their greatest breadth; their colour a dark-green. Both the leaves and the bark possess a very aromatic smell, perceived when they are broken or rubbed in the hand. The bark, besides, has a sharp biting taste, bearing some resemblance to pimento, and sweetish, which remains some time in the mouth after chewing it.

The thickness of this bark is proportioned to the size of the tree itself, so that we gathered some pieces above an inch in thickness. This bark consists of two coats closely united; the outer of an ash-colour, and the inner coat, when fresh cut, is of a dull white; but it soon turns reddish, and at last becomes of the colour of chocolate.

The seed resembles pepper, containing each four or five grains, black, longish, and semicircular; and their taste and flavour like that of the bark, but much more pungent and active. The seed forms clusters of five or six together.

We planted some young trees of this kind in the soil where we found them, of which we took the greatest care; but they all dried up when we got into the neighbourhood of the Equinoctial; so that, unless the seeds we brought home shall spring, we shall be disappointed in our hopes of transplanting this valuable tree in Spain.

The qualities of this bark show that it possesses a corroborating, antiscorbutic, stomachic, virtue. We made use of it as spice or pepper in our soup and broth, which not only gave it an agreeable flavour, but also an advantageous property. The water that filtrates commonly through the roots of this tree, we found acquired from it a digestive and laxative power: so that it is proper to be cautious in the use of this bark in cold climates, or in winter season; for, even in summer and in warm countries, so far from doing good, it might be hurtful, as it is too irritating.

In the moist and marshy places we found a kind of shrubs, or small trees, resembling the cypress, growing up extremely straight and even. They are covered with branches from the very ground: their greatest height from four to five yards; the greatest thickness of the stem, or trunk, ten to twelve inches. The leaves are like those of the cypress, and of the same colour, and only differ from them in being four-cornered. The fruit is small and black. When we saw them, they were dry and hollow, with a seminal powder within them. The taste of the leaves is excessively bitter, even surpassing that of broom.

In the woods is seen a species of palm, whose stem is about one yard high, and its thickness twenty inches. The branches are placed at the top, opposite to each other, in the manner of the date-tree, but never forming a cluster: the largest of them about one yard, having the leaves joined together, like fern or polipody; their colour bright-green, and of a disagreeable taste. We did not find any fruit on these shrubs, which abounded in the neighbourhood of streams of fresh water, and which, from their general resemblance, may be considered as belonging to the palms.

The *amarillo* (yellow tree or bush), or thorny tree, is in height from two to three yards; its leaves and branches all beset with prickles and thorns; the leaves and branches of a

dark-green on the outside, but within, and also in the stem, of
a deep-yellow. It produces a fruit like a mulberry when ripe,
of a feeble acid taste; but stronger when unripe, and possessing
the same virtues.

The only quadrupeds observed in this mountainous district,
are the dogs, of the same race with those of the Patagonians,
from whom, most probably, the natives have drawn them; and
a sort of deer, which we saw only at a distance, and therefore
cannot say precisely of what kind they were, as there are great
varieties in these animals.

Sarmiento, in his account of his Passage through this Strait,
mentions his having seen traces of tygers: on this we can only
say, that we met with nothing of that kind; neither does this
climate seem at all proper or suitable for such animals.

Much more numerous than the quadrupeds are the birds
which inhabit the woods; but the shortness of our time, and
the few opportunities we had of shooting or hunting, and the
difficulty of ascertaining the names of each kind, obliges us to
confine our remarks to those alone which most frequently presented themselves to our view.

It is a commonly-received opinion, that cold countries and
severe climates are but little abundant in birds; and that even
those they produce are not so beautiful, nor possess such variety
and brilliancy of colours, as those of the torrid zone: nevertheless, we saw in this part of the strait a kind of green ravens,
of the size of a pigeon, with some scarlet birds, like those of
Chili, of a very beautiful appearance.

More beautiful, however, is a little bird, about the size of a
sparrow, whose feathers are black as jet, and have a very
narrow golden stripe running lengthways, and a yellow beak.
This contrast of colours forms a whole of a very beautiful appearance.

The magpie is also met with, which differs little from that
of Spain.

Snipes or woodcocks (*becasinus*), are also very common, of
the same kind with those found in the Falkland Islands, and of
an exquisite flavour.

We also saw very frequently a kind of bird resembling our
blackbirds (*mulus*), but which is, without doubt, of a different
species.

What caused much astonishment to us, was to meet,
even on the mountains covered with snow, some small
birds, to which, on account of their resemblance to ours,
we called Magellanic swallows. Perhaps these are not, in

spite of appearances, of the same kind with those which, in autumn, retire from Europe, to enjoy a more temperate climate in their native country.

The singing and voice of these birds, and of all the other sorts, have nothing in them very agreeable; so that our ears were never much entertained by any sounds we heard to proceed from them.

The Indians presented to the commander, Don A. Cordova, a small humming-bird, dead and dried up, but with almost all its feathers, in every respect similar to those found in the hottest climates. It is difficult to explain whether that class of birds can exist in such a country, seemingly so opposite to their nature; but, as we saw only this single bird, and that one dead, we could not determine whether it had been driven by some accident to this part of America, where it perished, or whether the climate of 52° S. latitude be in reality so contrary to its nature as has been hitherto supposed.

Birds of prey are not wanting in this region, which live by the plunder and destruction of the weaker classes; but of them we are unable to furnish any proper information.

Much more numerous, and considerable in point of utility, are the aquatic birds. The geese, both the common and the royal, are here found in great abundance, and are of an excellent flavour.

Another kind of fowl, not less common, although their flesh smells a little of shell-fish, but which is nevertheless tolerable food, is found in the bays of the strait, and in the neighbourhood of the rivulets. They are smaller than the geese, their feathers black-and-white; the beak scarlet, and long. They are always seen in pairs; and, when pursued, utter a very singular and particular kind of whistle.

There is also a prodigious number of another kind of geese, called by our sailors *bustards*, because their flesh is abominable and insupportable: their feathers black-and-white, the neck long, the head of moderate size, the beak yellow, the tail very short; their flight not very rapid, and in general they fly in large flocks.

The flesh-birds, or bone-breakers, as they are called, abound in the strait; some of them extremely large.

The sea-gulls also are here in prodigious numbers, and of many different sorts. The most beautiful is one not bigger than a large turtle: its head is black, the whole body and wings of a dazzling whiteness, mixed with a few light black strokes; diamonds and rubies themselves do not equal the brilliancy of its eyes, round the pupils of which is a circle coloured like carmine, which adds infinitely to the beauty of this animal.

The penguins never fly; but so rapid is their course over the surface of the water, making use of their wings in place of oars, that they leave behind them a track in the water like a ship's wake; so that it is very difficult to take them: however, when on the shore, they very seldom escape. M. l'Abbé Perneti, in his very judicious History of the Falkland Islands, has treated at sufficient length of all these birds.

In all the strait we met with no poisonous animal nor troublesome insect; in which observation we agree with all other voyagers to this part of the globe. It is true that, within the woods, are found a few mosquitos, but they neither bite nor incommode with their buzzing noise; nor do they ever come out of the shade of the trees, no doubt, on account of the severity of the open air. There are also some butterflies and field-spiders: a number of beetles are also seen, but a little different from those to be found in Spain.

We have but little to say respecting the fish, which, contrary to our expectations, is by no means abundant in this strait; and only in the neighbourhood of the rivers are any found, which however, we must say, are of an excellent flavour. Neither with the net nor the rod did we catch more than four kinds: one called mullet, of all sizes, but the largest not exceeding six or eight pounds. In the Falkland Islands this fish is called *bacalao;* for, when it is cured, it is not inferior to that brought from Newfoundland. The next sort is the *espercuro,* which is not so common. The third is very small, of a reddish gelatinous quality, whose tribe we do not know. And, lastly, the fourth is the *pexe rey* (king-fish), some of which weighed about half a pound; well deserving the name they bear, if not on account of their size, at least from their delicious flavour, for, when fried, they are most exquisite.

We also saw some whales, porpoises, and sea-wolves or sea-lions; but it is needless here to say any thing concerning these cetaceous animals, already so well known, and which are precisely of the same kinds with those which frequent the seas and coasts of Spanish America.

But, to make up for the scarcity of fish, the shores of the strait abound with most excellent shell-fish. The *megillones* mussels, lempits, spouts, sea-snails or whelks, and sea hedgehog, constitute the chief nourishment of the Indians; but not of the Patagonians, who draw no part of their food from the sea; and were also the constant feast of our ship's company in the strait. The mussels, in particular, whose size is often five or six inches, are not inferior in flavour to the richest oysters; for which reason, to distinguish them from others, naturalists have given them the appellation of Magellanic mussels. In many of them are found pearls, produced, according to the general opinion, by a disease to which that shell-fish is subject. The lempits are of an uncommon largeness, and the inner part of the shell is a mother-of-pearl of great beauty; but they are neither so agreeable to the taste as the mussels, nor by far so easily digested.

In our nets we used to bring up also a number of *santozas* lobsters, and a kind of crab, of tolerable quality.

All these sorts of shell-fish feed, in general, on the juice of a marine plant called *cachiyuyo* or *cachiyullo*; but, by the naturalists who accompanied Cook in his first voyage to the South Seas, *fucus giganteus antarcticus*, as being peculiar to this southern hemisphere. The stalk of this plant extends to the surface of the water, being in length from fifteen to twenty feet. Cook indeed says, that there are some plants of sixty to seventy feet; but, in the course of our voyage, we met with none of such extraordinary length. The roots are attached to the rocks and stones in the bottom of the sea, and are of the same colour with the plant itself, which is of a dark-yellow, resembling that of dead leaves of trees when they begin to dry; the stalk is about the thickness of the finger; it discharges a mucilaginous slimy juice. From space to space are seen small longish bladders, of little thickness, filled with water, from which springs the leaf, about two or two and a half feet long, and in its greatest breadth from four to five inches. This leaf ends in a point, being shaped like a very sharp-pointed almond; it is not smooth on the surface, but neatly figured with longitudinal lines, a little raised above the root, so that, at a distance, they resemble the water-ribbons. From each root spring up five or six of these stalks or branches, so closely set together as often to cover entirely a large space of the sea, and so thick, that it is with the greatest difficulty that a boat can pass over them.

The sight of this plant indicates always a rocky or stony bottom; so that, if possible, navigators ought to avoid sailing near it, on account of the inequality of the depth of water constantly found where it grows. In many parts, vast quantities of cachiyuyo are found floating on the surface of the sea, torn from their roots by the tide and force of the wind; so that the shores of the strait are generally covered with it.

Such is the soil and climate, and such are the productions and animals, to be found in the Strait of Magellanes. This is the whole stock furnished by nature to the inhabitants, of whom it is now time to offer some information.

SECTION II.

Of the Inhabitants of the Strait of Magellan.

THE extreme moisture, and consequent unhealthiness, of this climate, is the cause of the low state of population in this strait. The inhabitants are composed of two races of men, different from each other in every respect; those who dwell in the plains, and those who inhabit the mountains. The latter, in particular, are in very small numbers; for, from Cape Negro to Cape Forward, we only saw one tribe, consisting of forty or fifty persons, who followed the frigate all the way to Cape Redondo; and we are even inclined to believe, that they do not extend to the westward of Cape Forward, as they are in all appearance the same individuals that were met by Bougainville and other French navigators, when they went to the strait to provide wood for their settlement in Falkland Islands. It is not so easy to ascertain the number of the other inhabitants of the Strait, since we only observed about seventy persons, but perhaps there are many more. The plain country at the east end is much better peopled; but, as there is so striking a difference between these two classes of Americans, we shall give separate accounts of each class.

The inhabitants of the plain country on the east and north part of the Strait of Magellan are the famous Patagonians, who, under so many different appearances and accounts, have furnished such ample room for the investigations and discussions of the literati of Europe, and have for so long a time kept up a high degree of uncertainty respecting the existence of a nation of giants; in favour of which notion they seemed to offer a most convincing argument.

When the elegant and judicious Dr. Robertson wrote his History of America, he remained undecided on this subject, on account of the great diversity of opinions received in the world; expressing his surprise, that, seeing all animals are found to arrive at their highest perfection only in the temperate climates of the earth, in which the necessaries for their sustenance and preservation are found in the greatest abundance, nature should have reserved for the ungrateful region of Magellan, and for a tribe of wandering savages, to exhibit the highest honour of the human race, distinguishing them by an increase of stature and a vigour of body superior to those of all other men.

We will not enter on discussions foreign to our purpose, respecting the long-disputed stature and strength of the people before the Flood; or whether, by a particular exertion of her powers, Nature has preserved in any corner of the globe a race of gigantic size; but merely content ourselves with asserting, that, in no respect whatever, can the Patagonians be called a nation of giants.

The many and careful measurements and observations of the officers of our frigate, exactly tallying with those made by Messrs. Carteret, Ulloa, and Bougainville, correspond precisely with the correct description left us of these men by the two Nodales, who passed this strait in 1618, viz. that the Patagonians were strong-limbed, well-personed men, *(unos hombres membrudos y a personados.)*

It is impossible to avoid taking notice of the heavy charge brought against the first Spanish navigators and travellers, of having given rise to the fable of the gigantic stature of the Patagonians, in order to render still more extraordinary the occurrences of their voyages, which, for that purpose, certainly needed nothing but a simple relation of the truth. A great number of authors, who have taken a share in this contest, all agree in affirming, that Magellanes Loaisa, the brothers Nodales, and even Sarmiento himself, all, with one accord, bear witness to the gigantic stature of the Patagonians. But we, who have had access to, and carefully examined, the original journals of these celebrated navigators, have found in them nothing in the least supporting such an opinion.

The Journal of Alvo, which alone remains of all those made in the expedition of Magellanes, does not even make mention of the Patagonians. Urduneta, in mentioning that he had intercourse with them, calls them *grandes y feos*, (large and ugly.) The MS. of Meri, although it treats particularly of these natives of South America, takes not the least notice of their stature. The same may be said of Camargo. Ladrilleros affirms, that the people of the strait are well-bodied men, stout and stately, and of great strength. The inhabitants of the mouth, towards the north, have large bodies, both men and women; the men being of great strength, and very alert and active. Sarmiento, in his first voyage, says only, that ten men could hardly hold one Patagonian; and, without expressing their size, calls them in general *giants*, saying, that, to the east end of the river San Juan de la Possession in Port Famine, begins the race of big men. In his second voyage, he adds, that the Indians of the strait were large and tall, and their captain or chief the biggest and tallest: and he always distinguishes the inhabitants of the north mouth by the epithet, the Great People; saying, that one he took on-board was of large limbs. Her-

nando Tomé, the Spanish seaman who was delivered by
Cavendish, in his passage through the strait, affirms, in his declaration, that the Indians with whom the colony of Sarmiento had
intercourse were of gigantic size; and, in another place, that
they were very bulky and ugly, or unshapely. We have already mentioned the opinion of the Nodales; so that, upon the
whole, it is evident, that not one of the first Spanish navigators
attributes to the Patagonians that prodigious stature, nor the
other extraordinary circumstances spread abroad to the world,
by authors who, even down to Seixas, in 1690, have written
multitudes of inventions and falsehoods.

The Italian Pagifeta, in the romance which he published, as
a History of Magellanes' Expedition, is the first who gave to
the Patagonians a stature of more than four yards; but, ab-
stracting from the little credit due to this author, on account of
the absurdities and falsehoods scattered over his work, in the
circumstance of their size he is so inconsistent, that, after having furnished them with heads of a monstrous size, he says
that Magellanes, amongst other presents, gave to one of them
his own cap, which the other immediately put on and wore,
although Magellanes was himself far from being a giant. So
fond was Pagifeta of such prodigies, that he has even planted
giants on the banks of the Rio de la Plata.

Maximilian Transilvanus, who in his work only translated
Pagifeta's book, repeats the same absurd story, and even em-
bellishes it with some improbabilities of his own invention.

As these two were the only works which had general circula-
tion in the world, they fell into the hands of authors of more
simplicity than discernment, such as Gonzalo Fernandez Oviedo,
who translated the above and many other fables into his General
and Natural History of the Indies; enlarging much on the
Patagonians, confiding in the information he received from the
clergyman Arizega, who, wantonly abusing his credulity, told
him many things which do not appear in the formal declaration
he afterwards made, and which are totally undeserving of credit: such as, that even a tall man could not reach with his
hand to the waist of a Patagonian; that these people devoured
a couple of pounds of raw flesh at a mouthful; that they drank
off six or seven *arrobas* (18 or 20 gallons) of water at a draught,
and other ridiculous exaggerations, to be seen in the above
history.

When this opinion came to be published by Oviedo, a co-
temporary author, it is not to be wondered at that it began to
gain credit, and was adopted by Gomara, Argensela, and other
writers, who, with excessive credulity, committed to writing

whatever they heard of these remote regions. Gomara not only copied Pagifeta, but added many other absurdities, collected from other quarters.

Nothing can be farther from our intention than to wish to bring into ridicule these historians of acknowledged merit ; we only mention these things, to show how easily they might be induced to give credit to the gigantic stature of the Patagonians. Whoever is at all acquainted with the writers of the sixteenth century, knows well that *that* was not the age of philosophy. Erudition, indeed, abounded in it ; but criticism arose in the following century. Notwithstanding, those authors who possessed a superior degree of judgment and discrimination, did not embrace that opinion, as was the case with Acosta ; who, although he repeats the stories of the giants said to have been found in various parts of America, does not mention the Patagonians as such, although it came within the limits of his history: and though he has allotted a separate chapter of his work to give an account of the Strait of Magellan. The famous Camoens likewise, who might, in his poem, with such propriety, have represented that nation as of prodigious stature, has, however, simply related the matter-of-fact.

It is also highly noticeable, that, of the great number of writers who gave credit to this fable, not one Spanish mariner or navigator is to be found, who, as an eye-witness, has given the smallest support to such an imposition.

The English traveller Cavendish is the first who, with other falsehoods, allowed 18 inches to the foot of the Patagonians, in which he is followed by Hawkins and Knivet. But those who have added the most to their bulk, were the Dutch navigators; and Sebaldus Veert even goes so far as to say, that they were of a height to fill men with horror; that they used to pluck up whole trees by the roots, &c. &c.: and he has been followed, but with some moderation, by Noort, Spilberg, and Le Maire. The adventurers of St. Maloes likewise bear testimony to the tallness of the Patagonians; but other travellers, and those undoubtedly of more credit, such as the English Winter and Narborough, the Dutch admiral S. Hermite, and the French M. Froger, have refuted these impostures; and also such as have made no mention at all of that people, are evidently against the vulgar opinion, as they never would have omitted a matter of such note and singularity. But, after all, the great stature of the Patagonians remained problematical, having many supporters on each side ; although the assertors of the gigantic size of that people were far from being agreed amongst themselves, fluctuating in their imaginary measurements from 10 to 13 feet; until the latest and repeated expeditions of the English, the French, and the Spaniards, in our own times, under the command of officers of distinction, judgment, and veracity, have for ever overturned and destroyed those fables and falsehoods, and reduced the Patagonians to their true and proper stature and appearance.

This however, it is true, has not been effected without some opposition; for, prior to the publication of the journal of Commodore Byron, and the Letter from Captain P. Carteret, in the Philosophical Transactions, a person who called himself an officer belonging to the Dolphin, Byron's ship, published his account, insisting much on the extraordinary size of the Patagonians, and other circumstances equally unfounded. And as this Journal, unfortunately, fell into the hands of Dr. Ortega, of Madrid, he translated it into Spanish; and, in his introduction, overpowered with this authority, and also by his own mistake in believing that almost all navigators that have passed the Strait of Magellan are of the some opinion, he finds himself obliged, in some measure, to conceal his own opinion in a matter of such nicety. We, however, who are determined not to conceal our real sentiments on the subject, have already shown what it is; and also how much the world was imposed on by that English author, who relates many things entirely omitted by his commander himself, in the very judicious account of his voyage he afterwards published.

As so many falsehoods have been circulated respecting these Patagonians, it shall be our business now to state the little certain information we have been able to collect concerning them; for we must observe, that, although we had various opportunities of intercourse with them from our first arrival at the strait, we are still unable to say much concerning their government, customs, and manner of life.

The Patagonians, so called by Magellanes, and not by Cavendish, as is said by the first editor of Byron's Voyage, are a collection of wandering savages, who occupy all that vast tract of country extending from the Rio de la Plata about S. latitude 37°, to the Strait of Magellan in S. latitude 52° 20'. Their most settled habitations are in the interior of the country; but, in the hunting-season, they approach the strait, where navigators have met with them.

Their stature, so much disputed, exceeds, in general, that of Europeans. Some of them being measured accurately, we found that the tallest did not exceed 7 feet 1¼ inches, Burgos measure; and that the common size was from 6½ *to* 7 *feet.** But even this height is not so striking as their corpulence, or rather bulkiness, some of them measuring 4 feet 4 inches round the breast; but their feet and hands are not in due proportion to their other parts. They all give evident signs of strength of body: they are full of flesh, but cannot properly be called fat. The size and tension of their muscles evinces their strength; and their figure, on the whole, is not disagreeable, although the head is large, even in proportion to the body: the face broad and flattish, the eyes lively, and the teeth extremely white, but too long. Their complexion, like that of other Americans, is *cetrino* (pale-yellow), or rather verging on a copper-colour. They wear thin black straight hair, tied on the top of the head with a piece of thong, or ribbon, brought round their forehead, having the head entirely uncovered. We observed some with beards, but which were neither thick nor long.

Their dress adds much to the effect of their size, being composed of a kind of cloak made of the skins of lamas or zorillos, arranged with some skill, with stripes of different colours in the inside. They wear it fastened round the waist, so that it covers them below the calf of the leg, letting that part commonly hang down which is intended for covering the shoulders; and when the cold, or other cause, induces them to put it over them, they hold the upper part of it with the hand, and so cover themselves entirely with this cloak.

Some also, besides this skin-cloak or mantle, wear *ponchones,* and breeches or drawers, of the same shape and sort with those worn by the Creoles of Chili and Buenos Ayres. The *poncho* is a piece of strong cloth striped with various colours, about three yards long and two broad, having an opening in the middle, made for the purpose of passing it over the head: a piece of dress extremely proper for riding on horseback, as it covers and defends the arms, at the same time leaving them in perfect liberty for any exertion. Some had ponchos made of the stuffs manufactured by our settlers in Buenos Ayres. The breeches, or drawers, are very like those worn in Europe; but their boots are very different, being formed of the skin of the legs of the

---

* The *vara*, or yard, of Burgos, the standard of Spain, contains 33·06132 inches, or two feet nine inches and one-sixteenth, English; the tallest Patagonian, therefore, did not exceed six feet six inches and one-third, English; and those of the common size were from five feet eleven inches and two-thirds to six feet five inches and one-seventh, English. It is, however, to be remembered, that *Spaniards* are not in general tall men, and that a *seaman* is seldom among the tallest of his countrymen: to *them*, therefore, the Patagonians might appear giants.—It may be as well to mention here, in explanation, (*vide* p. 12,) that six feet eleven inches and a half, Spanish, is about six feet four inches and a quarter, English.—*See the Table of Lineal Measures.*

horse, taken off whole, without cutting them open, and sewed up at one end.

There were, however, few Patagonians who enjoyed all these conveniences. The far greater number were almost naked, having only their skin-cloak, with a sort of leather purse hanging by a thong fixed round the waist, and fastened between the legs with one or two thongs, to the former round the waist.

With a piece of skin or leather fastened round the foot, they make a kind of shoe, and fix to it, behind, two little bits of timber, forming a sort of fork, which serve them for a spur; but they leave off this part of their dress when they have no intention to go on horseback, which, however, happens very seldom.

It is a very general practice among them to paint the face with white, black, and red, a kind of ornament contributing very little indeed to the agreeableness of their appearance.

Their equipage, or horse-furniture, consists of a kind of covering formed of several skins of lamas, one over the other, and rolled up a little both before and behind, so that at first sight they have some resemblance to a saddle; the whole fastened on with strong leather thongs or straps, instead of girths. The stirrup is formed with a piece of wood four inches long, supported at each end by a small thong, connected with another which is fastened above to the girth. The other parts of their furniture resemble entirely those used by the Indians of Buenos Ayres, with this difference alone, that the bit is made of very hard solid wood.

As the Patagonians have neither iron nor cordage, they supply their place with solid timber, and straps and thongs of skin or leather.

We saw one among them having a complete European saddle and bridle, but could not learn by what means he had acquired them.

Although we saw these people in troops of 300 or 400 together, yet we can give no information concerning their women, who never came near enough to permit our examination: only an officer, who was on shore in the Bay of St. Gregorio, assured us that their stature was somewhat shorter than that of the men, and that they differed very little from the men in their dress.

The children, even in their tender years, show that they are descended from parents of extraordinary size; and, by the largeness of their features, indicate to what they will arrive, when nature shall have attained its full vigour, and their members shall be properly developed.

As the Patagonians draw no part whatever of their nourishment from the sea, they have established their dwellings in the interior of the country, in the valleys near to some rivulet or pond of fresh water, and in the neighbourhood of some mountain, to shelter themselves from the fury of the winds, so that we could have no opportunity of seeing their manner of lodging and living, having only, now-and-then, on-board the frigate, had a distant view, from which we were not enabled to remark many particulars. Nevertheless, since we know that they lead a wandering life, like the Arabs, abandoning the tracts that fail spontaneously to furnish them food, it is natural to suppose that their huts are constructed without design or solidity. As a proof that they lead an unsettled wandering life, we can alledge, that, during our voyage, we met with the same tribe established in two different parts of the strait.

It is difficult to speak with certainty of their temper and dispositions, considering we had so few occasions of intercourse with them; only, we can affirm, that they are neither barbarous nor cruel, and that it is an atrocious injustice done to them by Cavendish and M. Gennes, who, in their journals, attribute to them the horrible practice of eating human flesh; and that, in this way, they destroyed and devoured the unhappy Spaniards of Sarmiento's colonies in the strait;—an imposture of which it is not easy to discover the origin, as the Spanish authors who have written concerning these settlements, are totally silent on such a subject. Their peaceable orderly behaviour ought not to be attributed to want of sense, courage, or spirit; for the objects of their most anxious desire were our arms,—a proof of their brave and warlike dispositions; so that, knowing their own superiority in strength over the other classes of Indians their neighbours, it is but natural to suppose, that they are not wanting in the means of gratifying their vengeance: but this does not contradict their general peaceable demeanour; for we never could discover in them any marks of evil intention in their dealings with us.

It has formerly been observed, that they used to leave their horses, arms, &c. on the shore, when they came on-board our ship, which proves the reciprocal good faith existing among them; and the frankness and readiness with which they put themselves unarmed into our hands, demonstrates evidently that, unconscious of treachery in themselves, they are not apt to suspect it in others.

We saw them entrusting to each other, on merely lying down on the beach, their most valuable articles, in order to embark, sure of finding them again, on their return to the shore; and so much are the rights of property respected among them, that the ribbons, presented to them by Byron, not being in sufficient quantity to be distributed to every one, those who did not receive any share of the present, showed no kind of dissatisfaction, nor once attempted to disturb the joy expressed by their more fortunate companions.

Nor is the idea of commerce a stranger to these men; at least, such as it was in the earliest times, before the multitude of commodities rendered it so complicate, as no longer to be confined to the simple barter or exchange of articles. Some of our officers exchanged swords and cutlasses with them for their furs, in which the Patagonians manifested the strictest honesty and sincerity. That they are extremely sober and temperate, is known from their constant refusal to taste wine or any other strong liquors, of whose pernicious effects we soon discovered that they were not ignorant. But this does not prevent them from requiring a large supply of food, proportioned to their uncommon bulk, to satisfy the demands of their appetite. They are also no strangers to the generous virtue of gratitude; for every time that our boat carried them back to the land, they always made signs for our people to wait, until they should go to procure some presents and refreshments for them.

They seem to acknowledge some kind of subordination among themselves; for the seaman Hernendez Thomé, who was brought off by Cavendish, relates, that a Patagonian fell into a violent passion, on being made to understand that Sarmiento was the captain or chief; and, giving himself several blows on the breast, asserted that he himself was the *capitano*. It was also observed, that the generality of one tribe paid obedience to one of more than ordinary size among them, intimating that he was their *capitano*,—a word whose signification they seemed perfectly to comprehend. It is, however, totally unknown how far this power extends, or whether the same persons always live in the same community; whether their numbers are great or small, or to what bounds they extend their peregrinations; what sort of religion, or if any, they profess; having on this head only remarked, that, before the sun went down, they constantly retired from the sea-shore to their habitations up the country, showing a sort of veneration for that beneficent celestial body.

It cannot be doubted, that the greater part of these Patagonians have frequent intercourse with the Spanish settlements in Buenos Ayres and Chili, particularly with those lately established on the Patagonian coast; for they all were well acquainted with the use of tobacco, which they begged of us with great earnestness, and showed themselves to be dextrous in the art of smoking. Besides, to be thoroughly convinced of this intercourse, it is easy to hear them pronounce many Spanish words, of whose meaning, however, they were generally ignorant; and to observe, that they possessed several articles of furniture and arms of Spanish manufacture.

The Patagonians possess most uncommon facility in repeating any word they hear pronounced, and even retain it in their memory. The seaman brought away by Cavendish, says, in his declaration, that he often heard them pronounce the words *Jesus Santa Maria*, looking up at the same time to heaven; and that they made the Spanish settlers understand that, up the country, there were other men with beards and boots, and other children like those of the settlers. This faculty of retaining words and phrases of different languages, has been constantly remarked and admired by all travellers in these seas. Captain Wallis mentions, that he taught some of them to repeat distinctly the English phrase "*Englishmen, come on shore;*" which, on his meeting with them many days afterwards, in another place, they correctly repeated. This power seems to proceed from their not having any harsh, nor indeed any peculiar, accent in their own language, from their acute hearing, and from their extreme volubility and pliableness of tongue, and other organs of speech. There is nothing either particularly harsh nor sweet in their language, which is very full of vowels, and its pronunciation somewhat guttural.

We proposed to several of them to carry them home with us to Spain, promising to bring them back again to their native country, but all answered, that they did not desire to leave their comrades; so that we did not deem it either reasonable or just in itself to take advantage of our superiority, to drag these men, against their will, from the bosom of their country and families,—to them so valuable and dear; especially as no other fruit could be reaped from such a step, than merely the satisfying idle curiosity with the sight of men whose stature exceeds that of the ordinary race of mortals, but which does not exceed, nor indeed even reach, that of particular individuals who have frequently been exhibited in Europe as preternatural productions of the powers of nature.

If ignorance of those things which it most imports man to know, and of the comforts and security of civilized life, which seem to be so congenial to man's nature, were not, in our opinion, insurmountable obstacles in the way of happiness, few men could be in a situation to render them more contented and happy than these Patagonians. They enjoy all the essential benefits of society, without being subjected to the infinite multitude of pains and uneasiness which too high a state of refinement never fails to produce. They enjoy strength and health, the fruit of their temperance; and they are ignorant of the baneful effects of luxury. They possess a wide field for the satisfaction of their limited desires, which are in proportion to the scanty and confined state of their ideas. As the soil they inhabit produces spontaneously their necessary food, they are relieved from numberless and various labours, passing their days in tranquillity, indolence, and repose, the love of which seems to be their ruling passion, and the certain result of all circumstances attending their state of life, and by no means that of natural stupidity, or inaptness for exertion, as many have asserted, but of which we discovered no proof: for the philosopher will not attribute to stupidity the high value they set on a string of glass beads, or other trifles of the same sort; but rather to the desire implanted by nature in the human mind, of ornamenting himself, that he may become more agreeable in the eyes of those around him; which desire, when well considered, is much more excusable in these Patagonians than in the European is that of diamonds and pearls, and other precious jewels, which are only acquired often by immense labours and dangers; whereas the Patagonian, without any toil or hazard, procures his ornaments when they fall in his way, by exchanging for them the surplus of those things which he already possesses, but which he does not immediately want.

The happy situation of these Patagonians is still greatly enhanced, in our opinion, when we compare them with their neighbours, the inhabitants of the mountainous parts of the strait, on whom we now proceed to make some remarks.

Of the Indians of the Strait of Magellanes.

The other class of inhabitants of this strait consists of a very small number of men, who, according to all accounts, can be compared only with the wretched beings who dwell on the western shores of New Holland.

Notwithstanding our long and continual intercourse with these natives, both in Port Famine and in Port Galan, we were not able to collect any certain information respecting the religion or civil constitution of the different tribes or families which we examined; for, instead of fixing their attention on the signs we directed to them expressive of our desires, they only, by way of answer, repeated back to us the very same signs and sounds; so that, after a very long conversation of this kind, both parties remained just as ignorant as at the beginning of it.*

Their manner of life being so little elevated above that of the brute creation, and their societies so limited in number, it was impossible for us to learn any thing more concerning them than what fell under our own observation,—that is, their figure, food, arms, navigation, and arts, if such we may call the manufacture of the few rude utensils they possess.

Before we arrived at Port Famine, our boat having been on-shore, brought off five of these Indians, found on the beach, whose nakedness, stupidity, and insupportable stench, filled us with equal horror and commiseration; for they seemed to live in the utmost wretchedness. When sent back to land, they joined their companions who had been left behind them; and, all together, followed the vessel all the way along the coast to that port.

It is impossible to describe how offensive and loathsome these men are from the filth of their persons as well as of their huts, constantly covered with the shells of shell-fish, and the remains of their ordinary food.

It seems to be beyond a doubt, that sometimes, though probably seldom, they have communication with the Patagonians; as was evident, from the perfect similarity of their dogs and lama skins, which they procure in exchange for some productions, but what we know not, peculiar to their country. However, their striking inferiority in bodily strength, as well as in mental powers, will naturally induce them to avoid much familiarity with their eastern neighbours, from whom they are totally different, but entirely resembling the natives of Tierra del Fuego.

They are of the ordinary size, rather inclining to middle stature; their limbs are well-proportioned, and they are very agile, notwithstanding they use very little exercise; their complexion pale-yellow, inclining to a copper-colour, but some of them darker than others. There is nothing either remarkably disgusting nor pleasing in their features. Their hair seems to be rather like that of horses and cattle than that of human creatures, which is probably occasioned by their having the

---

* *Note of Original.*—Our officers, who, for a space of two months and a half, kept up a continual intercourse with these Indians, notwithstanding their utmost exertions, were able to collect only the scanty hints contained in the following pages; and were not a little inclined to admire the facility with which other travellers have been able, in the course of only a few days, to acquire an intimate acquaintance with their customs, laws, religion, and even their language!

head constantly uncovered, so that, if it were properly taken care of, it would become fine and long: its colour is black. Some among them had beards; but they were very thin, and far from common.

The women are somewhat shorter than the men, and have nothing characteristic in their countenance; but evince an extreme care in covering what nature dictates to be concealed, as well as their breasts, which are generally large and hanging low down. So sharp and delicate is the voice of the women, as to differ from that of the men much more than is observed between the sexes in any other place with which we are acquainted.

The principal ornament of these Indians is a cap or bonnet of feathers, worn only by the eldest among the men; they also paint the face, legs, and 6ther limbs, with various stripes, white, red, and black, which only render their appearance more disagreeable; they take vast pains in this part of their ornament, as we could observe, particularly when they were to come on-board our frigate. The skin of a sea-wolf, or a seal, thrown over the shoulders, and descending as low as the middle of the leg, which they fasten to the waist with a cord made of the entrails of fish, is all they use for dress and to defend them from the weather, except a sort of apron made of feathers, which hangs before them: sometimes they wore on their feet a piece of the same skin, fastened and drawn round the ancle like a purse. The women wore this skin-cloak not only fastened round the waist, but they also brought it round under their arms, securing it round the neck, so as entirely to cover the bosom.

One part of dress, peculiar to women of all ages, is to wear round the wrist a sort of bracelet made of the entrails of fish; the same they also place round the small of the leg above the heel.

Both men and women have a cord tied round the head like a coronet, which, in some measure, confines and keeps up the hair. Round the neck some of them wore strings of beads, as they may be called, made of small shells tolerably well prepared; or else many turns of a small cord made in form of a necklace, made of entrails of fish.

Children of both sexes are commonly quite naked; and we could not help wondering at the great bulk of their bellies, which, however, return to their proper size, in proportion as they grow up. This appearance is perhaps owing to their never using any kind of bandages or swaddling clothes; for when their children are born, they are laid lightly on some skins of young seals, and the mothers commonly carry them about with them to all places, placing them in a kind of bag, which is itself placed within the skin which covers their shoulders. It is not uncommon to see some of these women with two children, one older than the other, situated in this manner, without being at all hindered, by this double though precious load, from performing the services to which they are destined.

There is no doubt that the principal food of these Indians is shell-fish, which is found in great abundance all over their shores; and the wandering unsettled life they lead, is caused by the necessity they are under of changing their place of abode, when the stock is consumed; that nature, providing for their wants, may have time again to produce those very shell-fish in the places where they are become too scarce to afford them an useful supply.

The deer which are met with in this region of the strait fall sometimes into the hands of these Indians, as we perceived, not only by their skins, but by pieces of flesh which could have belonged to no other animal. As they possess a great number of dogs, they probably use them in the hunting of deer; which, however, are not often seen; so that they are but seldom taken, as we observed that they seldom penetrated into the thickest of the woods. It is probable that the natives wait for these animals near the river-sides, whither they are used to repair, and there, with the help of their dogs, and with sticks and stones, are enabled to kill some of them; for their arrows and other arms are by no means proper for such sort of hunting. We could not be sure whether they eat this flesh raw, as they do the shell-fish, or if they put it on the fire; but it is probable that they do not take that trouble, as they have neither instruments nor vessels fit for preparing their food.

The different feathers we saw in their possession showed that the birds do not always escape their darts; and, as they are very dexterous in the use of the bow and sling, it is most probable, that they make use of one or other of these ways to obtain their end; however, they do not seem much addicted to this practice, as we never saw but a few birds together in their hands: and we are of opinion, they do not eat their flesh, as we never observed in their hovels any marks of such food.

Much more common than hunting of any kind, with them, is fishing, in which they are more dexterous. Although we are ignorant of their mode of fishing, we know they are much given to it; and they even came twice with a good quantity of fresh fish to sell on-board our vessel.

They possess neither nets nor fishing-hooks; only we remarked that, at low-water, they used to fix sharp-pointed stakes in certain parts of the shore, forming a kind of fish-weir; we are not sure, however, that by this method they catch the fish. They also take along with them, in their canoes, several long sharp-pointed stakes or poles, with which we imagined they struck and killed their prey, having on the end a sort of bait, tied on with a piece of cord. It was impossible to make them comprehend that we wished to learn their mode of fishing; neither did we ever happen to witness their contrivances and management for that purpose.

We imagine that, when these Indians pass over from the continent to the Tierra del Fuego, their chief object is the catching of tunnies, seals, and whales, which seldom visit the northern shores of the strait. They eat raw the flesh of these most indigestible animals, even when putrid and stinking, and of the fat make a sort of oil, with which they continually anoint themselves; on which account, it is easy to be sensible of their approach a long way off, as their smell is insupportable.

They also eat, and for that purpose keep in their huts and canoes baskets filled with, wild fruits. When they saw our men eating wild parsley, they pointed out to them several other plants and roots on which they feed, roasting them like potatoes. Of all the things presented to them on-board, fat and suet, or tallow alone, seemed really to please their palate: they rejected bread, oil, and vinegar, and never could be induced to taste wine.

Their dogs also live on shell and other fish, and herbs, which proves that the change of climate, and the necessity of obtaining some kind of food, has entirely altered the appetite of these animals; they however with their appetites have not changed their proper good qualities, and are still the most faithful guardians of their masters.

The habitations of these Indians consist in some miserable huts of a round form, composed of branches of trees, with the thick end fastened in the ground, and the small one upwards and inclining over the middle of the hut, connected together with ropes made of rushes or grass. The circumference of their largest huts is not above eight yards, and the height two. The only opening or door is low, and in width about the eighth part of the circumference. When they reside in their huts, they cover them all round with seal-skins, just as they come off the animal, for they know not how to dress them for any purpose, leaving uncovered the middle of the upper part, to let out the smoke of the fire they constantly keep up, and which is placed in the centre of the hut, round which they place boards or benches covered with straw or rather dried grass, which serve at once for chairs and beds. When their fire is extinguished, they kindle it again with a flint; and, for tinder, make use of the feathers of birds.

Their furniture and utensils consist of several skins of seals, deer, and a few of the lama or guanacos, which they certainly obtain from the Patagonians, as none of these animals are found in this part of the strait; a number of baskets made of rushes, and others of a species of *sparto;* several jars of one foot and one and a half foot in circumference, made of the same bark with which they construct their canoes, having the bottoms sewed in, in the same way: they are worked with some skill and neatness, and capable of holding any liquid, without its running through. They likewise possess some small bags made of skins or of guts of fish, in which they keep their different powders for painting themselves, where we may notice that scarlet is their favourite colour; the strings of beads or necklaces made of small shells or bones; the stones from which they extract fire; and other little articles of this nature.

Such are the contemptible goods of these ill-fated mortals, which they carry about with them, when they change their place of abode.

Their boats or canoes are made of the bark of the tree producing resin, whose greatest thickness does not exceed one inch. They are formed of three pieces; one in the middle, forming the bottom and the keel, and the other two pieces the sides. Their patience and application are admirable in stripping the bark of those trees, having for the purpose no other instrument than a flint somewhat shaped and sharpened, with which they make an incision round the trunk at each end, and then another lengthways, to join them; afterwards, with vast patience and management, they strip off the whole bark of the tree in one entire piece, of the proper length for the intended canoe, which, in some, including the bend of the middle-piece, which forms the stem and stern as well as the keel and bottom, is from 30 to 32 feet; and the true length of this frail boat, when finished, is from 24 to 26 feet; the greatest breadth four feet, and the depth from two to three feet.

In order to make this bark acquire the proper curvature and shape, and lose what it has on the tree, they lay it on the ground, with the inside downwards, and on each end place a heap of stones, leaving it so for two or three days, in which time it dries and becomes fit for being employed. Then they place the side-pieces almost perpendicular to that in the middl , joining them together with seams of dry rushes, and caulki or filling up the interstices with dried grass and clay or mud, as much as possible to prevent the entrance of the water. To give some strength and resistance to the sides, they lay across the canoe pieces of wood resembling pipe-staves, one by the side of another, all along the length of the canoe, giving it the shape of a semiellipsoid, and make the ledge or gunwale of two strong poles, well joined together at each end, and in them are fixed the ends of the cross-pieces, which serve as ribs or timbers; the whole being tied and sowed together with rushes; placing also crosswise from time to time, some pieces of wood, that answer the purpose of benches or thwarts to sit on.

When the canoe is in this state, they line almost the whole of the inside with pieces of the same bark, about one foot broad, laid across, and having the ends made fast in the gunwale on each side. In order to give these pieces the requisite bend, they heat them by the fire, and, when they are half dried, apply them to their proper situations. Besides this, they form a kind of floor from the fourth part from stem and stern, placing it about half a foot from the bottom, leaving an opening in the middle, to throw out the water. This sort of floor consists of boards laid lengthwise over others placed crosswise; and, as well as all the rest of the canoe, is covered with bark.

Such is the construction of their boats, which, although but rudely wrought, do not fail to cost these Indians much time and labour, for want of instruments and tools proper for such works, which, indeed, are the only ones in which they show any ability. They were not ignorant of the advantages to be drawn from knives, hatchets, and nails; and soon showed us that they preferred these tools to any other contrivance. Some of them procured such instruments, and even tried to imitate them with pieces of pipe-staves.

Many of these canoes are capable of containing nine or ten Indians: they are moved along with a sort of paddles, and rowing them is the ordinary employment of the women. When they enter on a long voyage, which is always either in a calm or with a fair wind, they set up a pole as a mast in the bow of the canoe, and across it another like a yard, having fixed to it the skin of a seal, and keep the lower parts of it steady with their hands: and this scanty sail relieves them from the fatigue of rowing. In the middle of the canoe are some stones, with heaps of shells and sand, on which sort of hearth they make their fire, keeping it up with branches and sticks.

Belonging to each canoe also are several jars, such as those before described, which serve to throw out the water which makes its way into it. Besides, each has several ropes or cables made of rush and esparto, of various lengths and thickness, much like the small ones made of the same materials in Spain.

It appears impossible that in such frail vessels they should undertake voyages along and across the strait, in a climate so inconstant, and so subject to sudden changes from calms to furious squalls. Nevertheless it is certain, that they very frequently cross the channel, and even make long courses in the strait, of which we had sundry proofs; for instance, that of the family that followed us from Cape Negro to Cape Redondo; and that in the mouth of the channel of S. Geronimo we met with many of those Indians we had before seen in Port Galan. In these canoes they also transport their whole furniture, and other property, when they migrate from one part of the coast to another.

The boldness with which they run such hazards, may perhaps be attributed to their intimate acquaintance with the strait; but which, on many occasions, cannot prevent them from falling victims to their rashness and imprudence.

Their arms consist of the bow and arrows; the bow being roughly made of wood, with a cord of entrails of fish, with which they give it the requisite curvature. The arrow is made of a smooth stick, of the length of two or three feet, having at one end a piece of flint worked down and shaped like a heart; and, at the other end, two small portions of feathers, fastened to it with a very fine cord. No very considerable efforts ought to be expected from this weapon: however, they are extremely dexterous in the use of it, and we observed that they fixed the arrow easily in a tree, the flint separating itself from the wooden part.

The sling is applied to two very different purposes: one is to discharge stones, and the other to fasten their skin mantle round the waist. The stone is situated in a piece of skin or leather, and the cord or string is, as usual, of the entrails of fish.

They also carry sometimes a stick of two and a half feet in length, and as thick as the finger, having a stone or flint, of the same shape as that of the arrow, fixed at the end of it, two inches long, and of proportional thickness: this they employ two javelin, throwing it with the hand.

On many of these Indians we also observed a sort of dagger or dirk made of bone, and very sharp, of various shapes. They tie them fast to a pole or staff two yards long, and use them, we believe, to kill seals, whales, &c. as they have some resemblance to our harpoons; and, when near their prey, they may inflict no less mortal wounds.

Amongst the natives we saw at Port Galan, were some who possessed small pieces of iron fixed in wooden handles, in imitation of our hatchets, chisels, and augres, which, without doubt, have remained in their hands since the last visits of English and French travellers, on whose tools they set a high value, on account of the great assistance they furnished in facilitating their labours.

The skill and dexterity with which they manage their different weapons, and the scars visible on many of these natives, show that, on some occasions, they have quarrels and wars amongst themselves; but we can affirm that they are not continually engaged in hostilities against their neighbours, and that those of Tierra del Fuego are not at all times enemies to those of the continent, as we have seen them reciprocally paying visits to each other; and therefore it is only on some transient occasions, that their differences break out, which, however, are soon brought to a proper termination.

It is extremely difficult to ascertain the numbers composing one tribe or family, or to discover whether, when they unite together to the number of 60 or 70, they consider themselves to be all related, and to form but one single society.

We only remarked, that every eight or ten live together in one hut; and that, although many more may reside at the same time in one place, still each family is composed of the above number of persons, and that each takes charge of providing its proper food and fire, of the education of the children, and of their own hut and canoe.

To the women is entrusted the care of gathering shell-fish, fruits, and herbs, for the support of the family and inhabitants of the hut to which they belong, as also to keep up the stock of wood and water for daily use; to keep the canoe dry and clean, on which account they are often up to the middle in the water: they likewise row when they are in the canoe; and, lastly, have the entire charge of rearing their children,—in doing which, they employ the most anxious care to preserve them from diseases and accidents common to the state of infancy, which, among those Indians, seems to be of shorter duration than in civilized societies.

The men, far from taking any share in the labours of their women, reserve themselves entirely for certain other occupations; such as constructing of canoes, building of huts, making of arms, and employing them in hunting and fishing. But their labours are neither so severe nor so continued as those of the poor women; you therefore see them, for the most part, sitting on their hams, their favourite attitude, round the fire, or stretched out along the beach, whilst the females are engaged in unremitting toils and anxiety for the support of the family. What a difference between the practice of this country and that of other parts of his Catholic Majesty's dominions! Perhaps in both there is an excess. Although these Indians seem to set but little value on their women, beholding them with much indifference, and being little if ever visited by the terrible passion of jealousy, yet they did not much relish the attentions paid to them by any of our people.

We could obtain no information respecting their marriages; whether one woman is common to two or three men, or what degree of consanguinity they observe in forming those connections: only we were struck with the vast disproportion between the numbers of both sexes; for in every family or tribe that we examined, there seemed to be at least three men to one woman. Our notions of this nation are too imperfect to enable us to explain this disproportion, which must certainly be one chief cause of the smallness of their numbers.

Their language is difficult, and to all our ship's company was totally unintelligible: it seems not to be abundant in expression, and the pronunciation is almost entirely guttural, so that the same word uttered by different Indians seemed to become almost a different word. On this account, we never came to comprehend any part of what they said, nor even to be able to repeat their sounds; whereas they, on the contrary, repeated with great ease and readiness whatever they heard us say. A favourite word, which they pronounced almost every instant, was *pecheri*, which we explain as equivalent to friend. M. de Bougainville has given the name of *pecheri* to these Indians, merely from the circumstance of their incessantly using that word in their conversation.

Their dispositions seemed to be peaceable and not ill inclined: they never attempted to purloin any thing from us, notwithstanding that the sight of our instruments, utensils, and tools, must have created in them a vehement desire to possess them, even at any hazard; but perhaps this moderation and orderly behaviour, on their part, proceeded more from a sense of their own inferiority, when compared with us, than from any moral principle or sense of the injustice of appropriating to themselves the property of others.

To this quiet demeanour of these poor men, and the extreme care bestowed by our commander, D. A. de Cordova, we must attribute the perfect harmony and good understanding which uninterruptedly prevailed between them and us, during the whole time we were amongst them. We must, nevertheless, acknowledge that, among themselves, we never discovered the smallest dispute, nor the faintest symptoms of anger or revenge. All these virtues may, however, be the natural effect of the supreme indolence and laziness which predominate in their hearts; and which must, no doubt, account for the inconsiderable advances they have made in society and civilization.

Whether or not this evil is more to be considered than the amicable enjoyment of peace and concord, which is its consequence, is a question which we leave to the decision of the philosopher.

Curiosity, which seems to be an essential and universal characteristic of the human race, has not hitherto made any progress in the hearts of the inhabitants of the Strait of Magellan.* Nothing we presented to their view excited in them the feeblest admiration or wonder, nor even a desire to be better acquainted with it. In order to be capable of admiring the productions of art, it is indispensable to possess, at least, elementary ideas of these productions; but these men beheld the most delicate and complicated works of men as they do the laws and phenomena of nature itself; and made no difference, in their estimation, between the composition of the mast of our frigate, and the natural trees which spring up in their forests. The haughty European, who, after exposing himself to numberless dangers to arrive at their abode, thinks he humbles himself much even in holding intercourse with these unfortunate beings, must be not a little mortified at perceiving the supreme indifference with which they survey the recent productions of his industry and talents.

---

* The length of the Strait of Magellan is nearly 107 leagues. All former navigators have allowed it 12 or 16 leagues more; but, in our support we can say, that Cook, although he did not pass through the strait, observed this excess, by the difference of the longitude which he observed of its mouth west from the Strait of La Maire, and that of this place, as recorded in Anson's Voyage, calculated from Cape de las Virgines.